Banishing Math Anxiety

Sheila Tobias
Math Anxiety Consultant, Tucson, Arizona

with

Victor I. Piercey
Ferris State University

Kendall Hunt
publishing company

Book Team
Chairman and Chief Executive Officer: Mark C. Falb
President and Chief Operating Officer: Chad M. Chandlee
Vice President, Higher Education: David L. Tart
Director of Publishing Partnerships: Paul B. Carty
Senior Developmental Editor: Lynnette M. Rogers
Vice President, Operations: Timothy J. Beitzel
Assistant Vice President, Production Services: Christine E. O'Brien
Senior Production Editor: Carrie Maro
Permissions Editor: Patricia Schissel
Cover Designer: Mallory Blondin

Cover Image © 2012 Shutterstock, Inc.

www.kendallhunt.com
Send all inquiries to:
4050 Westmark Drive
Dubuque, IA 52004-1840

Contents

About the Authors

Sheila Tobias

The term "math anxiety" is attributed to Sheila Tobias, whose book *Overcoming Math Anxiety* set the agenda for taking students' emotional blocks to learning math into account at all levels of mathematics instruction. She is also the author of *Succeed with Math, Breaking the Science Barrier,* and *They're Not Dumb, They're Different,* which deals with learning difficulties in the physical sciences. She makes her home in Tucson, Arizona.

Victor Piercey

Victor Piercey joined Sheila Tobias in preparing *Banishing Math Anxiety* as her sometime partner in math-anxiety workshops. With a PhD in Mathematics from the University of Arizona, Victor brings a wide range of experience in teaching math from the elementary to the college level. His teaching extends to remedial college courses and courses for elementary education students, where he works hard to help students deal with their math anxiety. He recently joined the faculty at Ferris State University as Assistant Professor of Mathematics and lives with his wife in Big Rapids, Michigan.

Acknowledgments

The authors would like to graciously thank Nancy Angle of the Mesa State College Mathematics Department for her thoughtful comments and suggestions of the manuscript as a whole.

The authors would also like to acknowledge Erin Dokter of the University of Arizona's Office of Instruction and Assessment for contributing to the initial conceptualization of Short Take #5 and Chapters 7 and 8.

How to Use This Book

Note to the Student

This book is meant to ease your way into college mathematics. It doesn't matter whether or not you are currently in a math class. We think it's useful to review some of the math you have already learned and maybe forgotten and to experience some success. If you gave up on yourself and learning math in your prior schooling, here's a chance to start fresh.

There are two messages that run through this book—messages we don't believe you have been exposed to in the past. The first is that active learning—asking questions, consulting your instructor and other students out of class, using the tutoring center, and making sure you're in a study group—helps you take charge of your own learning.

Our second message is that math learning may begin in the classroom, but it doesn't end there. In the last two chapters, we take you on a tour of the real world of pre-employment testing, math applications on the job, and managing your debt, your taxes, and (eventually) your investments.

We don't believe that some students can and other students can't do math. And when you finish this book—if we have succeeded in our purposes—you won't believe that either.

Note to the Teacher

This is not a textbook. It's a book about math anxiety that will benefit students most if their instructor reads the book in its entirety before starting to work with students. What it isn't is a pre-college math textbook, although mathematical content is substantial. Nor is it a collection of do's and don'ts—neither for the instructor nor for the students.

This book, rather, is issue oriented: using resources, deciphering notation, mathematics appreciation, uses of math in the real world. Some of the chapters are very original, such as "Tools of the Trade," which brings the student to an appreciation of graphing calculators without presuming that they have ever seen one, or formally teaching calculator techniques. The chapter on problem solving is intended to encourage creative exploration instead of some step-by-step approach that is supposed to work every time.

In addition to eight chapters, we have included five "Short Takes" meant to elaborate and further instruct on the topics covered. Teachers might create exercises out of the Short Takes. Short Take 4 provides some calculator explorations. Short Take 5 describes a process that enables a group of students to instruct one another.

For further reading, the authors recommend Sheila Tobias's *Overcoming Math Anxiety*, which was written to assist adults whose fear of mathematics limits their career options and their self-worth.

Chapter One

Making Math Work for You

There are two myths about mathematics. Myth One: that college-level mathematics is hard, harder than other subjects, and harder than it has to be. Myth Two: that without college math, you can have a well-paying job and enjoy a wide range of activities. We don't believe these myths, and when you finish this book, you won't believe them either.

In the pages that follow we're going to help you banish math anxiety from your life by offering you new ways of:

- Thinking about mathematics
- Doing mathematics
- Studying mathematics
- Talking the language of mathematics
- Eventually figuring out how to learn mathematics on your own

Mathematics is no longer just an entry-level requirement for engineering, the physical sciences, and statistics. Welcome to our twenty-first-century world. The basic principles of math, along with computers, have become part of almost every area of work. Its logic is applied to everyday problems people have to solve. This is a big change from the days when most occupations were virtually math free. Today, even if college-level algebra and statistics are not in the job description, they will give you an edge on the competition later on for promotion into management or work in more interesting technical areas.

The reason is this: Mathematics gives us a way of *describing relationships* that would otherwise be too complicated to make sense of in any other way. Try explaining to a customer, without drawing a graph, why even if a gas guzzler is cheaper to buy at the outset, it may be more expensive to run over the long haul.

Math Anxiety

Where does it come from?

We have taught or interviewed hundreds of math-anxious college students. What they have in common is this: They can all remember the classroom where they began to doubt that they had what it takes to learn math. Maybe it was the "girls don't do math" message, or the fact that no one in their family ever chose a math field, or that no one in their ethnic group ever became engineers. Others had teachers who sold them the either-or myth: that they would either be "good with numbers" *or* "good with words," but that they couldn't do well at both. Then there was the image of the geeky mathematician. Because our American culture doesn't consider mathematicians as role models, why try to be like them? Besides, math seemed like dreary, hard work, and never any fun.

These are all misconceptions. Let's take them one at a time. First, if there still are few females and members of minority groups in the top tiers of working mathematicians and scientists, it is not because they are missing a "math gene" but because there have been social and institutional barriers for them to overcome that are only slowly disappearing.[1] Second, while some writers don't like math and some mathematicians don't like to write, there is no evidence that writing ability and mathematics ability are mutually exclusive. On the contrary, college admissions counselors tell us that the student who shows high capability in *both* the mathematical and verbal sections of college-entrance exams is more likely to succeed overall in college than the student who has a score that is severely skewed toward math or verbal abilities. Finally, although elementary mathematics may appear to be repetitive, there are skills that must be practiced, like playing a musical instrument or practicing a basketball throw, if you are to get to the creative parts later on.

A legitimate complaint of math students is the *style* of the mathematics classroom. Students object to the fact that there is little opportunity for them to talk—except to ask questions—and no opportunity for discussion or even debate! Many prefer English or social studies classes both because they can participate and because there is not so much pressure on tests to find a single right answer. Mathematics does depend on right answers, but it can also be experienced as a series of discoveries that we make for ourselves. Too often than not, however, math is presented as a fixed set of rules to be digested whole and without dispute, which discourages many active students from learning.

And then there are the tests.

"I used to panic so much about timed tests," one college student told her math-anxiety counselor in a college that offers a math-anxiety clinic, "the only thing I really learned in all my years of elementary schooling was how to do quick subtraction and short division." This student wasted precious test-taking time looking at the clock, counting the minutes she had left for the test (that's the subtraction part) and then dividing the total minutes by the number of questions she had left to do. Of course, she was getting more and more anxious as she went through this exercise.

Few people can think clearly and well with a clock ticking away. It's hard to perform at the blackboard with thirty sets of eyes watching you. No one likes a subject that is presented rigidly and uncompromisingly. And most people do not do well when they are scared. We believe that "math inability" may not be the

[1] See, e.g., Sheila Tobias and Susan Haag, "Math Anxiety and Gender: A Review of the Literature—1970–2010," *The Encyclopedia of Diversity*, Sage Publications, 2012.

result of a *failure of intellect* but, rather, of a *failure of nerve*, an idea that led to the establishment of math-anxiety clinics around the nation.[2]

How does math anxiety stop you from succeeding?

Think of your brain as a three-part system with an input area, a memory bank, and some kind of understanding and recall pathways connecting the two.

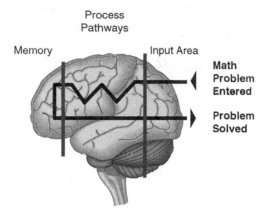

Process
Pathways

Memory   Input Area

Math Problem Entered

Problem Solved

© 2012 Kendall Hunt Publishing Company

Imagine you're in math class or alone doing your homework or studying the text. If you're feeling good about your math skills, or even just good about yourself, you will start by trying to make sense of the wording of the problem. What is being asked for? What am I being told? And, in time, you'll be able to call up from memory the right formula or approach. What you will be doing—in terms of the schematic above—is moving back and forth along the process pathways of your brain until you realize exactly what you have to do to solve the problem. Anytime you get stuck—if your mind is free to travel along those pathways comfortably—you will return to your memory bank or back to the wording of the problem, or refer to your textbook. You might be tempted to draw a diagram, or put some hypothetical numbers into the problem to make it more concrete. But whatever you choose to do, you will be *busy*, moving along the pathways of your brain, activating your memory, trying out solutions, checking your answer—busy.

Now suppose your memory bank is intact and your understanding and recall skills are well developed, but every time you look at some new mathematical material or problem, your emotions interrupt the pathways. You can't think. You

---

[2] There's still some debate about this. Some researchers find children with an inherited or genetic "dyscalculia" (5 to 8 percent of the school-age population) having "lower ability [than other students of the same age] to grasp and compare basic number quantities." Our research suggests otherwise: Dyscalculia as a measurable brain dysfunction is far rarer than dyslexia (inability to learn to read). Otherwise fully normal good learners have trouble with math because of the ways it's taught, societal attitudes, and an "old-century" view that one can lead a prosperous, fulfilling life without mastering some level of mathematics. See Sarah D. Sparks, "Study Helps Pinpoint Math Disability," *Education Week,* June 21, 2011.

panic. You tell yourself: "This is just the kind of problem I can *never* solve." You feel the tension that comes from time pressure or the uncertainty that comes from lack of confidence. What might your brain system look like then?

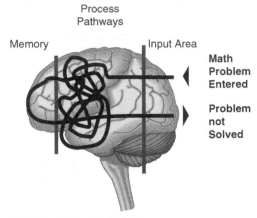

Your process pathways have become cluttered by emotions. There is an inability to think, but not because the "hardware" of your brain is inadequate. The input, memory, process pathways, and recall systems are just as good as they were before. But, because the pathways have been blocked by bad thoughts, you cannot remember. You lose confidence because you can't remember. You can't even *think*, no less analyze the problem. You may even doubt your own intelligence. But, in fact, the only reason you cannot work is that your feelings have created too much "noise" in your brain. Soon, your pencil (or your hands on the keyboard if you're doing your homework online) stops moving. Your brain freezes up. You can't work, you assume, because you can't think. But in fact it's just the reverse: *You can't think because you've stopped working!*

Managing Math Anxiety Means Taking Back Control

Active thinking

"Thinking" in mathematics involves doing. Did you ever notice that your math teacher can hardly talk to you about mathematics without writing on a dry-erase board or marking up a PowerPoint slide? Stories are told about mathematicians having lunch together who fill up their paper napkins and sometimes even the tablecloth with calculations, diagrams, and drawings while talking about their work.

They know something you don't know yet: Only by trying new avenues of thought, putting down first one idea (expression, equation) and then another, turning a diagram over and over in your mind, doing calculations, checking procedures, and checking answers can you learn math and solve problems. When you freeze, your emotions have made you stop. When you unfreeze your mind and spirit, you can move forward. The essence of doing math is not to stop, but to *keep going*.

From Self-monitoring to Self-mastery

College students who manage their math anxiety start by recognizing when panic starts, what form it takes (each individual is different), and how to un-panic systematically. How can anyone do all of that at once? The essence of math-anxiety therapy is self-monitoring.

One tool for self-monitoring is the divided-page exercise:

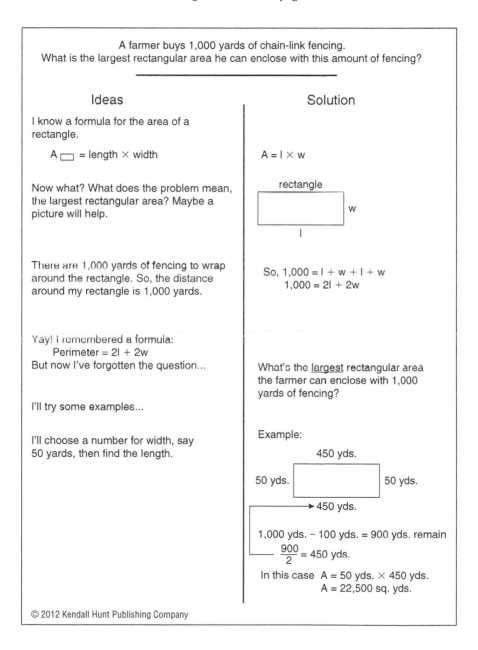

A farmer buys 1,000 yards of chain-link fencing.
What is the largest rectangular area he can enclose with this amount of fencing?

Ideas

I know a formula for the area of a rectangle.

A ▭ = length × width

Now what? What does the problem mean, the largest rectangular area? Maybe a picture will help.

There are 1,000 yards of fencing to wrap around the rectangle. So, the distance around my rectangle is 1,000 yards.

Yay! I remembered a formula:
    Perimeter = 2l + 2w
But now I've forgotten the question...

I'll try some examples...

I'll choose a number for width, say 50 yards, then find the length.

Solution

A = l × w

rectangle

w

l

So, 1,000 = l + w + l + w
1,000 = 2l + 2w

What's the largest rectangular area the farmer can enclose with 1,000 yards of fencing?

Example:

450 yds.

50 yds.   50 yds.

450 yds.

1,000 yds. – 100 yds. = 900 yds. remain
$\frac{900}{2}$ = 450 yds.

In this case A = 50 yds. × 450 yds.
A = 22,500 sq. yds.

Make yourself a divided page as in the diagram. Draw a line down the center of the page. On one side record your feelings and thoughts, however random and seemingly unconnected they may be. On the other side of the page, write your math notes, your calculations, and/or problem-solving steps.

You may not be able to capture your feelings and thoughts the first few times you try this exercise. But, eventually, you will learn to "tune in" to what you say to yourself when under stress. Once you've identified these pathways to anxiety you have a better chance to get rid of the emotional static that is getting in your way. To take one example: Suppose you wrote, as so many math-anxious students do when they are first given the divided-page exercise:

"This is just the kind of problem I could never solve."

The next step in your effort to free yourself from this prison of panic is to ask yourself:

"What specifically is making this problem difficult for me? And what can I do to make it easier for myself?"

"Maybe I need to draw a picture."

"Maybe I need to put in simpler numbers to get a 'feel' for the answer."

"Maybe I need to ask a well-framed question of my classmate or the teacher."

Instead of panicking, pick one thing to do and do it. If that doesn't work, try one of the other ideas.

Giving yourself permission

The purpose of the divided-page exercise is to give yourself permission to explore your own confusion and to find out what is making the problem or the new material seem more difficult than it really has to be. As you become more and more familiar with your own *learning style* and *learning pitfalls*, you will become more adept at avoiding potholes in the road. You'll soon notice that when you're busy filling out the left-hand side of the divided page, even if you haven't written anything on the right-hand side, you're not freezing up; you're still working.

This is the real point of the divided-page exercise: to keep you working when you're mathematically "stuck." Even when the right side of the page is blank because you can't figure out the calculations or remember a particular formula, you will be writing notes and comments—that is, you'll be keeping going. Just thinking about why you're having trouble clears some of the static from your mind, and as you observe yourself at work, you are using profound analytic powers—powers you may think you don't have when it comes to doing math!

Your ability to analyze your own resistance will become a source of insight into solving the math problem itself. Sometimes, students who use the divided-page exercise tell us that the right-side/left-side page jottings become so entwined that when they finally solved the problem, they hardly noticed they were working on it. They thought that they were still working on themselves.

Here's another example:

A car is driven 5,000 miles. Its five tires (one is the spare) are rotated regularly and frequently (every few miles). How many miles will any one tire have traveled on the road by the end of the trip?

On the left-hand side of the page, one student writes: "I know there's a formula for solving this problem, I even think I once knew it. But I've forgotten it and I don't think I can remember it in time to solve the problem. So, what to do? I'll try drawing a diagram—construct a diagram and plug in some numbers. Maybe the formula will pop up." It did.

From another student's left-hand jottings: "I keep coming up with 1,000 miles. But I sense that's wrong because 1,000 is too few miles for any one tire to have been on the road for that long of a trip. But maybe 1,000 miles is the right answer to another question. What could that other question be?"

And, from a third student: "What's confusing me is the word 'rotated.' "Rotated" seems to have two meanings: 'turning on the road,' and 'changing position on the car.' Could it mean both?"

See further discussion of this problem in Chapter 6.

Self-mastery

At this point, the extra time it takes to fill up the left-hand side of the divided page begins to pay off. Writing things down frees you from the almost paralyzing effort of staring at a problem or a page of text. The truth is this: *Thinking* in mathematics involves *doing*. That's why writing down seemingly unconnected feelings and thoughts breaks the tension and the sense of isolation. You are not alone. You are at least "talking" constructively to yourself. And in time, you will notice that the divided page provides a way to tune into your intuition and your common sense.

Best of all, the tuning-in process teaches you about your own learning style when it comes to approaching math. Some students don't feel secure, no matter what kind of problem they are trying to master, unless they draw a picture or a diagram. Others need to rephrase the question in their own words. Still others want to put numbers into algebraic equations to get a *feel* for the answer. When you've finished Chapter 4, you may find yourself needing to get comfortable with an equation by first sketching a graph.

Students who are successful in math are not necessarily smarter than the rest of us, but that they know themselves very well. They use that self-knowledge to anticipate the difficulties they are going to have and what kinds of questions and actions will give them the power and confidence to continue. They know when to skim a math text and when to focus on a particular paragraph or illustration. They are seldom bored because they are busy. They don't quit because they recognized long ago that progress in mathematics involves making a little headway, one step at a time. They don't judge themselves harshly when answers don't come out right. They are patient, tenacious, and don't expect to learn or do mathematics very fast.

Other Ways of Judging Math Competence

If we could pass two new "laws" in the testing of mathematics, the first would be to eliminate timed testing. No one but an airline pilot or a surgeon needs to get any kind of numerical answer fast. The rest of us (not in school) have plenty of time to study a problem, look up a table, calculate, and check our answers. The second new law would be to have students graded on "competencies" and not just the one right answer. Some of these competencies might be:

- How well they approach a problem to be solved (whether or not they get the right answer)
- How many different ways they can solve the problem
- How well they write a paragraph about what makes that particular problem mathematically interesting

The advantage of testing competencies in such a way is that it reduces the pressure and rewards thinking about a problem before and after it's been solved. And it is mathematical imagination, in the form of intuition and interesting analogies, that is the source of much invention and creativity in math.

It is not surprising (to us) that many students do not learn the value of imagination in their math classes. An algebra textbook states a problem and immediately refers to the relevant formula. Textbook authors and teachers don't talk to students about *their* many trials and errors the first time they tried to solve the problem. We get a false sense that they, the teachers, or they, the "good" students, do math instantly and with ease. What beginning students of math really need to know is how to start to work a problem when what to do first doesn't pop into their heads—even more, what to do when they don't know what to do.

Summing Up

No single book can transform anxiety or indifference into unbridled enthusiasm for mathematics. Our goal is more modest. The reverse of math anxiety is not expertise or mastery, at least not at the beginning. It is, rather, the willingness to learn the math you need, when you need it. Once you get to that stage, you will be able to function comfortably with mathematics in a work environment. You will not hesitate to ask for a quick "hint" from a colleague or for a link to a resource online any more than you would ask a local for directions.

Do you have to *think* like a mathematician in order to *do* math? We don't think so. If you can derive the formulas you have to work with, of course you will feel more secure. If you can make connections among mathematical principles, or better yet, link the math you've learned with applications in the world outside of school, you will have a better grasp of the whole.

As you lose your fear of mathematics, we hope you will allow yourself a sense of awe, for mathematics is a wondrous way of imposing order on the universe. Remember, it took thousands of years for the very best minds, working over time, to create the mathematical ideas ordinary college students learn today. That legacy should not be a burden to us, but should be taken for what it is: a gift! And above all it should be accepted, not avoided or left unexplored.

Chapter Two

Using Your Resources

Once you sign up for a class in mathematics, you will have resources to draw on, besides yourself, resources that you may not be aware of. First, there are human resources: your instructor, your classmates, and the college's tutorial center. Second, you have your textbook. Reading a math textbook requires very different reading habits than most students are accustomed to. For this reason, much of this chapter is devoted to techniques that can help you make the most of this resource. Finally, there are online resources that professionals use. In Chapter 5, we will introduce you to Wikipedia, WolframAlpha, and Khan Academy. When you discover how to use all of these resources, you will find that the power to learn is available to you in ways you never knew before.

Human Resources

Your instructor

There are good and bad instructors. Some are "good" for other students but just don't connect with the way you think. But either way, your instructor is a resource that you should make the most of. The real revelation you should have after you finish reading this chapter is that your instructor is not your only source of information and feedback, but just *one of many* resources available to you!

Nevertheless your instructor matters. Think about the qualities that you like in math teachers and those that do not work for your personality or learning style. Write these qualities down as a list. Then, talk to other students about the teachers in the math department at your school to find out who may or may not work well with you. Some teachers may be willing to meet with you before classes begin.

One challenge you may face in choosing an instructor is that your course schedule may simply read *staff* as the assigned instructor for every math course. Not to worry—it usually doesn't take much digging to find out who is assigned to teach which courses. Find the main office for the math department, where somebody may be able to help you.

Once you choose your instructor and classes begin, the way to get the most out of your instructor is to visit him or her during office hours. Your instructor will have designated time throughout the week when you can drop by the office to talk about anything you want. If you have conflicts with all of his or her office hours, you can make an appointment with your instructor at an alternate time.

Here's one survival strategy: Make it a point to attend office hours within the first two weeks of class in order to introduce yourself. This will be an opportunity for you to get to know your instructor, find out what his or her expectations are, and maybe even get a clue as to whether the course is at the right level for you. This will also give your instructor an opportunity to get to know you. Giving your instructor a face and a personality to attach to the name and student number on his or her roster will help to build a personal relationship, the bedrock for one-on-one instruction.

Once the course is settled, try to visit office hours routinely to go over anything that you may need help with. This could include a homework problem you are stuck on, an example from class that you didn't understand, or something from the textbook that was unclear even after you read and reread it. Always walk into office hours with an agenda—a list of tasks you would like your instructor's help in accomplishing—and a list of specific questions. Don't be embarrassed about this. As a college student you want to present yourself as a serious pre-professional. And this is the way professionals operate. Take notes during office hours, and don't be afraid to ask permission to write on the board in the office!

Taking this kind of initiative is a great strategy to make the most out of your instructor as a resource. And in addition, that relationship might lead to recommendation letters for internships, jobs, or graduate school.

Your classmates

Do not think of your classmates as competition but as fellow travelers on a journey to math competence. In fact, your classmates are another resource that can help you learn. In every math class, you can find a small group of students whose schedules will allow you to meet regularly. Set up study groups to meet in order to work on homework or prepare for exams. You will be surprised at how powerful peer learning is!

Tutors

Most colleges have organized free tutoring centers for all levels of mathematics. This is another resource you can call on for help. At large state universities, these tutors are typically graduate students who are assisting or even teaching the same courses that you may be taking elsewhere. Find out where the tutoring center is as well as its schedule. Drop by to find somebody that you click with. Don't be afraid to be selective. Your tutor should speak a level of English you can understand. Your tutor should be patient, willing to meet your misunderstanding halfway. If the relationship doesn't work out, look for another tutor.

Your Textbook

Reading a mathematics text, whether the text is online or in your hand, is not like reading any other kind of book. In books on other subjects, clarity often is achieved through repetition: using different words to restate a single idea; slowing the pace; using a spiral kind of organization that keeps coming back to the same idea at different levels; using topic and summary sentences to nail down what a paragraph contains; and foreshadowing the point to be made later on. Introductory and concluding sections are meant to be helpful to the reader.

Don't expect your math textbook to be like a history textbook. A history textbook tells a *narrative*, a story. The contents in a history book are designed to support that story and are organized so that it should be clear how the different chapters fit together as a whole. Not so with mathematics. A math textbook is organized into chapters and the chapters are organized into sections. Each section covers a different topic. Although the author(s) of the textbook may intend for an individual chapter or clusters of chapters to tell a kind-of story, that narrative will not be clear to the learner. Talk with your instructor and your study-mates about this. Ask them if they see a way to put all of these topics together, and why some sections come before others. These questions can lead you to one of the highest levels of thinking, *synthesis*, the assembly of disparate ideas into a single notion.

First, you have to slow down. Even speed readers can't speed read math. In mathematical writing, clarity is achieved by constructing very precise sentences without any extra words.

Second, in most mathematics textbooks, you will find summaries of key ideas at the ends of chapters. You may also find these summaries at the end of individual sections. These summaries can be very helpful if you are trying to tie several ideas together in order to create a narrative.

Third, a mathematics text contains diagrams, equations, and other kinds of illustrative material that you must not skim. You must study everything on the page, including tables, graphs, and drawings. Graphs in particular have to be taken apart visually to figure out their meaning. Special attention must be paid to mathematical expressions, which are the "vocabulary" of mathematics. In the Short-Take following this chapter called "Marking up a Math Text," you will see how some students make sense of a page of mathematics. Their techniques may help you find an approach of your own.

Finally, every chapter in a math textbook includes examples that illustrate the sentences and the words. Those examples are meant to be worked out because (like cooking from a recipe) you will only understand the information when you apply it to a problem.

Active reading, or writing as you read

Math textbook authors and the teachers who assign those books tend to assume that students will read their textbooks with pencil in hand, that they will go back and forth between the sentences, the illustrations, and the examples, and even look ahead to the homework problems, as they read. Properly done, reading mathematics is an activity! The meaning of a set of statements is only partially given by the text. The rest of the meaning has to be *constructed by the reader*. The problem is that no one may have taught you how to do this.

One reading habit some instructors encourage students to develop is keeping a running glossary of math terms in a journal. As you read through your textbook and review your class notes, when you come across vocabulary terms such as *equation*, *function*, *set*, or *element*, write the term and definition in your glossary. You should play a little bit with the terms by trying to write down examples. When you find examples that clarify the meaning of these definitions for you, include them in the glossary entry. You may also want to add to your glossary new notation, symbols used by mathematicians as shorthand. Keep in mind that your glossary should be a resource for you to read later. You want to write it in a manner that will remind you of what you learned when you first made the entry.

In the rest of this chapter, we will guide you toward reading a math textbook actively, section by section. Because you may be reading your textbook online, we will also provide advice on using computer-generated instructional materials. Finally, we will provide some tips regarding the use of textbook problem sets to prepare for exams.

Let's start with what might be for newcomers to mathematics a useful analogy.

Kit-building

Kit-building is an activity that illustrates one way to approach mathematical writing. Even if you have never constructed anything from a kit (or cooked a dish from a written recipe), you can well imagine how you would read the instructions that come with a kit. After a swift scan of the entire instruction booklet to make sure you have all the necessary parts and tools you would turn to the kit you want to assemble or the computer program you want to work with. The instructions make little sense until you start putting the pieces together.

It won't take you long to realize that the instruction booklet alone (or the software manual if you are trying to learn a new computer program) is insufficient. You must continually try out each new step or procedure. The instruction sheet gets you going, but here's the point of the analogy: It is in the figuring out and doing that the real learning takes place. That's true not only when building a model or mastering a new computer program but also when learning mathematics or science. What you have created in the kit-building activity is a three-way relationship between yourself, the thing you're working on, and what you are reading about the thing you are working on. If you and the instruction booklet were alone without the object to work on as you read, you would probably not understand much.

Applied to math, the analogy with kit-building works like this: Experienced and successful mathematics students *create tasks for themselves* as they read a mathematics text. After looking at one example, they cover it up and try to reconstruct it from memory. They then try to think of other examples that would fit the concept being introduced. They don't just *look over* the illustrative problems. They work them out; every one of them. They think of similar problems using the same principles or techniques, and try to solve them. In short, they treat reading a page in a math textbook as a set of *activities* in which they are going to participate.

Styles of Reading Math

Reading DOWN

Reading down—from the introduction to the examples—is one way to approach a math text (whether printed or online). The indented sentences that follow could have been taken from a textbook.

> **Proportions:** *Ratios are particularly useful in solving many consumer problems and also problems in science and business. Especially useful will be the case where we have two ratios set equal to each other.*

It would be good at this point to stop reading and try to think of an example— $\frac{2}{3} = \frac{4}{6}$ is one example of two ratios set equal to each other. But more useful, as we will see, is finding other examples at this stage of your reading. Once you have some comfort with this new term, you should copy the word and definition, along with some examples, into your glossary.

The textbook might continue:

> *Whenever we have this equality between two ratios, we say we have a proportion.*

You've just been given a mathematical definition. But not the kind you'll master if you memorize. Memorizing the words in a math textbook doesn't last long, but thinking of some examples will lock the definition into your memory. So, after reading the previous sentence, it's wise to start writing down examples of other ratios that are or may be equal to one another. Then stop and look over your examples and you'll see for yourself what a proportion really is. To test your understanding at this point, you might ask yourself:

- Can a proportion exist among three ratios (e.g., $\frac{1}{2} = \frac{2}{4} = \frac{3}{6}$)?
- Can a proportion exist if its ratios are themselves ratios?

You are now actively reading. If you are by yourself, you may appear to be talking to yourself. If you are studying with a classmate, all the better—you can both talk yourselves through the text.

The next sentence in this typical text provides an important new piece of information:

> *There are four members in a proportion, and if any three of them are known, it is possible to find the missing member.*

Here's a new word, *member*. We know that in common use, a member is a part of a group or organization. Is the word being used in that way here? Don't forget to add an entry for the word *member* into your glossary, because the mathematical use of the word may not be the same as how you would use it in casual conversation.

You should not be discouraged at this point if reading just a few sentences in a math text seems to involve so much work. A little extra work as you read new material in a math text will save you work (and confusion) later on. Math writing is densely packed. One way to "unpack" math sentences is to read them aloud and with a friend or a study partner to avoid possible misunderstandings. But even with active reading, only when you've done the problems at the end of the chapter can you say you've really "read" the chapter.

Reading UP

"Reading down" goes from an overview of the subject to the examples and the problems. Yet, many successful math students tell us that they do not read their texts this way at all, but instead prefer to "read up." Rather than starting with the explanatory paragraphs at the beginning of a chapter, they *begin* by looking at the mathematics itself: the examples, the sample problems, and the solutions to the problem sets. That's because these students have learned from experience that the words used in the textbook are confusing until even a careful reader works the examples.

Suppose we had completely skipped the narrative part of the text we just discussed and had proceeded directly to the examples. Try to verbalize the following example:

$$\frac{2}{a} = \frac{4}{7}$$

Now look at the sample solution and try to see where the author is attempting to take you:

$$4a = 14$$
$$a = \frac{14}{4}$$
$$a = \frac{7}{2}$$

Ask yourself why the author did each of these steps and in this particular order. If you cannot answer these questions, you should go back to the text and look for the sentence that explains the rule. Then, and only then, should you proceed to the next sample problem.

A variation on "reading up" is to start even farther down the text with the problem sets you'll be asked to do at home on your own. Let's face it, the only way you and your math teacher can be certain you have mastered the mathematical principles presented in the textbook is by how well (or poorly) you do the problems. So some students start with the homework problems, trying to work them out on the basis of what they know already, and only when they encounter difficulty do they go back to the discussion in the text and the examples.

In other words, they do not read any more of the text than they absolutely have to in order to make sense of each problem. They are not *reading* the math text at all, certainly not the way they would read history or literature. They are using the text as a reference, like a dictionary or a Google search.

This reading technique lends itself very well to online mathematics materials. If your teacher assigns online homework, chances are your software will give you several options when you are stuck on a problem. One of these clicks will take you to the portion of the text, presented online, where the concepts related to your problem can be found. All of this back-and-forth activity is even more efficient than when you are turning pages in a textbook. It takes place with the click of the mouse.

Online Textbook Resources

It is becoming increasingly common for college math courses to include an online component, most of which involves some homework done online. But the online course companions may include other features, such as an electronic copy of the textbook and visualizations. Then, there are the online resources professionals use, which we will introduce in Chapter 5.

When you are told that that there is an online component to your math course, you may feel heightened anxiety if you are a little uncomfortable with computers. Be reassured. You can master your online mathematics program because of the resources you have: your classmates, your instructor, your textbook, and technical support. Some of your classmates may be familiar with the online materials from prior classes. Your instructor may hand out step-by-step instructions for getting started on the program. He or she may even go through it with you in class. And there should be instructions bundled with your textbook. In addition, most programs feature 24-hour-a-day technical support staff who can be called on (or chatted with online) with any questions or for troubleshooting. Because deadlines usually come very quickly, you should contact technical support immediately whenever you have a problem with the functioning of your program.

Once you are signed up for your online components, take a brief tour of your materials. The software may include a tutorial to walk you through its features. Typically, the features include an electronic copy of the textbook, video lessons, animated lessons, online homework/tests/quizzes, and a grade book that is kept up to date (with your grades, not those of your fellow students). Many of these programs also feature personalized study plans and calendars, along with extra problems for extra practice. Don't exit your first session until you have figured out where your homework, quizzes, tests, and, above all, *deadlines* will be posted. If your system does not create a calendar for you, you should keep one for yourself whether on your computer, in a PDA, or via pencil and paper.

Using the online textbook

Online textbooks that come as part of online homework packages are well suited to the student who is inclined to reading up—that is, who wants to start with the homework problem and only go back to the text and the examples when in need of help. When the student starts having an issue with a problem, he or she can click one of several buttons to have content delivered. One option will be to jump to the part of the textbook that will help with the problem. Another option may be to see a video presentation related to the problem. A third option is to view a similar example worked out. A fourth is to be given the answer, work backward, and then try a similar problem on your own.

All of these choices may seem overwhelming. How you approach them may depend on how your homework is arranged. If you have a limited number of attempts, it may not be a bad idea to try some similar problems before you commit to an answer on the given problem. Ultimately, the process that works best for you is the one you will keep using.

Using Your Textbook to Prepare for Exams

When preparing for exams, most students spend more time solving problems out of the textbook than they do reviewing the text. After all, solving problems is what you will be asked to do on your tests.

One pitfall that you can avoid when you are working textbook problems is relying too much on back-of-the-book answers for self-assurance. When you are taking the exam, you will not be able to check the back of the book to assure yourself that you are solving the problems correctly, and the anxiety that this could cause will snowball.

In order to avoid this situation, try to mimic the exam environment while preparing. Select a handful of problems and work them out in a quiet room within the prescribed time limit. Only when you are finished should you check the answers in the back of the book to assess your progress.

When selecting problems to practice with, keep in mind that during the actual exam it may not be obvious what specific topic(s) a problem is testing. Review problems found after the summaries at the end of chapters may be helpful, as it will frequently be unclear what section a given review problem refers to. Try to write some of your own problems as well! Exchange problems with members of your study group and try to solve them. This type of practice can lead to greater confidence in your ability to take exams.

Developing *Yourself* as a Resource

The focus of this chapter notwithstanding, in some important ways mathematics textbooks and online materials are not meant to be "read" at all. As we have illustrated, words often get in the way of the mathematics. Its own notation is far better suited to expressing mathematical statements than sentences. The textbook can be thought of as a kind of scaffolding that is no longer needed once the structure of your understanding is in place. That's why some textbooks have few words and sentences and why many successful math students don't go back to the narrative in their textbooks when they review for an exam. They only review the problems.

Mathematics instructors know this and will often select a textbook or a set of online resources for the quality of its examples and the challenge of its homework problems. They rely on the text for *pacing*, as a way of communicating to students how far and how quickly they are supposed to be moving through the course. Yet, for those who are not used to a course of study where the problem sets matter more than the narrative, it is worth spending some time getting your instructor to teach you how to make best use of the resources provided.

Eventually the "best use" of the resources available to you in a math class will be uniquely your own design.

Short Take 1

Marking Up a Mathematics Text

We've all been taught to "take notes" and even highlight facts, examples, and conclusions from textbooks in other subjects. But nobody ever teaches us how to take notes from a mathematics text. Whether reading a text down, up, or from the problem sets, it is important to write notes, either in the margins or in your notebook. We want our notes to be so clear and exact that we can re-create our thoughts as we review them at a later date.

We've demonstrated in Chapter 1 how the divided-page exercise is a way of gaining control over your feelings. In the same way, making detailed notes helps you gain control over your understanding of any new material.

We have selected examples of marked-up textual material from an experienced math student. The segments are taken from a Kendall Hunt textbook.[1] We don't expect that you will be comfortable marking up your book at the beginning, nor that you will select examples and comments as useful as those of the student who marked up these pages. We are reprinting her notes in order to give you a sense of what active reading is all about.

Example 1

The first example comes from a chapter on linear equations and inequalities in one variable. Notice how the student makes sense of a statement that only has letters by putting in her own numbers. She chooses a to be equal to 2, and b to be equal to 3, and restates the algebra using these numbers. She ends up solving her own problem to check how well she understands the example.

She also translates some of the language into her own words, writes down connections to other topics, and asks herself questions. When she can't answer her own question, she makes a note to bring that question to class. Note also that she draws boxes around important rules. This way, the rules jump out when she is studying for tests.

---

[1] Lee C. Land, *College Algebra* (Revised Printing), Kendall Hunt Publishing Company, 2009.

C. Linear Inequality in One Variable:

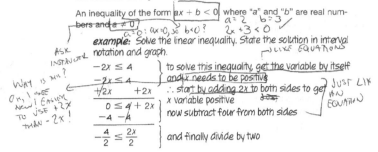

An inequality of the form $ax + b < 0$ where "a" and "b" are real numbers and $a \neq 0$.

handwritten: $a = 2$ $b = 3$
$a \neq 0$: $ax \neq 0$, so $b < 0$? $2x + 3 < 0$ ✓

example: Solve the linear inequality. State the solution in interval
notation and graph.

handwritten: ASK INSTRUCTOR — LIKE EQUATIONS

$$-2x \leq 4$$

to solve this inequality, get the variable by itself
and x needs to be positive

handwritten: WHY IS THIS? OH, I SEE. NOW I EASIER TO USE +2X THAN -2X!

$$\frac{-2x \leq 4}{+2x \qquad +2x}$$

∴ start by adding 2x to both sides to get

handwritten: JUST LIKE AN EQUATION

$$0 \leq 4 + 2x$$

x variable positive

$$\frac{-4 \quad -4}{}$$

now subtract four from both sides

$$-\frac{4}{2} \leq \frac{2x}{2}$$

and finally divide by two

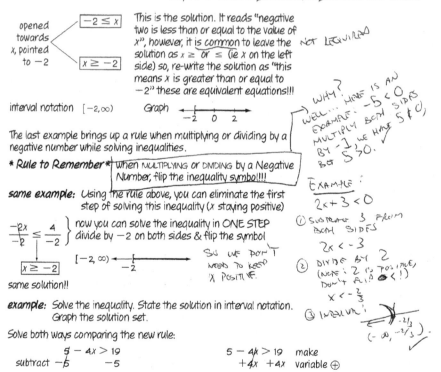

opened towards x, pointed to -2

$$-2 \leq x$$

$$x \geq -2$$

This is the solution. It reads "negative
two is less than or equal to the value of
x", however, it is common to leave the
solution as $x \geq$ or $\leq$ (ie x on the left
side) so, re-write the solution as "this
means x is greater than or equal to
-2" these are equivalent equations!!!

handwritten: NOT REQUIRED

interval notation $[-2, \infty)$ Graph

handwritten: WHY? WELL... HERE IS AN EXAMPLE: $-5 < 0$ MULTIPLY BOTH SIDES BY -1 WE HAVE $5 \sharp 0$, BUT $5 > 0$.

The last example brings up a rule when multiplying or dividing by a
negative number while solving inequalities.

* Rule to Remember * When MULTIPLYING or DIVIDING by a Negative
Number, flip the inequality symbol!!!!

same example: Using the rule above, you can eliminate the first
step of solving this inequality (x staying positive)

handwritten: EXAMPLE:
$2x + 3 < 0$
① SUBTRACT 3 FROM BOTH SIDES
$2x < -3$

$$\frac{-2x}{-2} \leq \frac{4}{-2}$$

now you can solve the inequality in ONE STEP
divide by -2 on both sides & flip the symbol

handwritten: ② DIVIDE BY 2 (NOTE: 2 IS POSITIVE, DON'T FLIP $\oslash < !$)
$x < -\frac{3}{2}$

$$x \geq -2$$

$[-2, \infty)$

handwritten: SO WE DON'T NEED TO KEEP X POSITIVE.

same solution!!

handwritten: ③ INTERVAL!
$(-\infty, -\frac{2}{3})$ ✓

example: Solve the inequality. State the solution in interval notation.
Graph the solution set.

Solve both ways comparing the new rule:

$$5 - 4x > 19$$
subtract $-5 \qquad -5$

$$5 - 4x > 19$$
$$+4x \quad +4x$$ make variable ⊕

Example 2

Our second selection comes from the same chapter in the same book. Our student is now dealing only with an example, so there is not much explanation. She makes it a point to understand each step before going on to the next one. For example, when she doesn't remember the "distributive property," she looks it up in the same book using the index or the table of contents. Then she writes down the page number of the explanation, so when she's studying for her test, she can go back.

Section 1.1. Linear Equations in One Variable

Another type of Multi–Step Equation is an equation with variables on both sides.

4. **Equations with Variables on Both Sides** when solving multi-stepped equations, the first goal is to get all the variables to one side and combine all like terms. This rule works for any equation with multiple steps.

example: Solve $-2(x + 17) = 13 - x$

Step 1: Get all terms with "x" to one side. To do this, eliminate the parentheses by <u>distributing</u> -2. · ? *See pg. 19*

[handwritten] $a(b+c) = ab + ac$
$a = -2, b = x, c = 17$
$-2(x+17) = -2(x) + (-2)(17)$
$= -2x - 34$ ✓

$$-2(x + 17) = 13 - x$$
$$-2x - 34 = 13 - x$$

[handwritten margin note] LIKE 1 & 2 STEP EQUATIONS SEE PG. 22

Step 2: Now get the variable on one side. This can be done two ways, *i.* add 2x to both sides, which keeps the variable positive; *ii.* add x to both sides, which keeps variable on left side. This will add an extra step at the end.

[handwritten] $2x + x = (-2+1)x \cdot 1x$ ✓
$-x - y$

i.
$$-2x - 34 = 13 - x$$
$$+2x \qquad\qquad +2x$$
$$0x - 34 = 13 + x$$
[handwritten] $-x + 2x = (-1+2)x = 1x = x$

ii.
$$-2x - 34 = 13 - x$$
$$+x \qquad\qquad +x$$
$$-x - 34 = 13 \quad 0x$$ ✓

Step 3: i. $-34 = 13 + x$ ii. $-x - 34 = 13$

to finish this form, only one-step is required. Subtract 13 from both sides.

[handwritten] $-34 - 13 = ?$
* $34 + 13 = 47$ ✓
$-13 - 34 = 47$

$$-34 = 13 + x$$
$$-13 \quad -13$$ ✓
$$-47 = x$$

∴ $\boxed{x = -47}$

* if done correctly, both answers should match

to finish this form, two steps are required, add 34 to both sides and divide by -1.

$$-x - 34 = 13$$
$$+34 \quad +34$$
$$\frac{-x}{-1} = \frac{47}{-1}$$ ✓

∴ $\boxed{x = -47}$ ✓

Step 4: With more complicated equations, the fourth step should ALWAYS be to check your work. You check your work by placing the solution into the original equation.

$$-2(x + 17) = 13 - x \qquad x = -47$$
[handwritten] OK ✓ $-2(-47 + 17) = 13 - (-47)$

"Marking Up" an Online Textbook

With more and more materials online, how do you "mark up" an online text? What you can do is either take notes in a separate notebook or print out selected pages to mark them up. If you are printing, that's a lot of paper! Even if you don't print many pages, it is probably a good idea to get a three-ring binder and a hole puncher in order to store what you've marked up in the sequence in which the topics are covered in class. Eventually, somebody may invent a "stylus" to mark up a textbook on your screen. But until then, taking notes in a notebook or on printed pages is what we recommend.

Short Take 2

Some Study Problems and Suggestions

How can you tell when you have studied enough? How do you know if you have gone over a chapter as many times as you need to? One good sign is that you are getting the right answers to most of the homework problems as you go through them, and that you seem to understand what is going wrong with those you miss. Second, notice your feelings as you work. If you are alert and confident, chances are you have mastered most of the material. Third, look back over the section on which you have been working. If it all looks easy now and you wonder why you had so much trouble at first, you have succeeded.

If you're not getting the right answers, if you don't feel clearheaded or focused, and the material is not getting easier, try to diagnose precisely what your problem might be so that you can take additional steps. One of the following might describe your situation. If so, try the suggestion given.

Problem: You can't understand the material as it is written in the textbook even after going over the sentences many times.

Suggestion: Try reading up.

Problem: You read and think you understand. But when you turn to the homework problems you discover that you didn't understand the material after all. This may be because the text is giving you rules and generalizations but not the details you need to do the problems.

Suggestion: First, look back at the general rules and relate various parts of those rules to the parts of the problem you don't understand. Then ask for help. When you go for help, be prepared to ask specific questions. (That's where having mastered the names of the elements in mathematical expressions or the types of graphic representations really helps.) Bring along your attempted work and keep the pencil in *your* hand during the tutoring session. This is a good general rule for all subjects, because if the tutor is doing the writing, then the tutor is doing the thinking.

Problem: You read, complete the homework problems, and do well on the quizzes, but you cannot seem to integrate all of the bits and pieces you learned when preparing for an in-class exam.

Suggestion: You may have to alter the way you do your homework. You may want to schedule one study session to review the lesson past, one to learn the current lesson, and one to preview possible exam questions—with a group, not on your own. Successful math students prepare for exams by redoing problems—not just *some* of them, *all* of them. They even invent new problems to solve, just to see if they really understand what they have been studying. Ask yourself why a certain approach was taken for a given problem, and see if you can alter the problem so that a different approach will be required for a solution.

Group Work: Studying with a group is a good way to reduce the isolation everyone experiences in learning math. Best of all, it provides a setting where it's essential to *talk* mathematics. Psychologists tell us we learn most efficiently when we employ all five of our senses (Mathematicians add another one: the imagination.) In fact, expressing new and difficult mathematical ideas orally is a good exercise in learning to think more deeply about what you are doing in math. Even if you aren't accustomed to working with groups, most tasks in the work world are accomplished in teams, and it is never too early to learn to handle a group dynamic.

Chapter Three

Deciphering the Code

Mathematics has its own vocabulary. In addition to words, mathematics uses its own notation, symbols that stand for more complicated ideas. Some of these elements are familiar, such as numbers, letters, and the arithmetical operations signs like $+$ and $-$. Some are less so, as in algebra, where mathematical notation also includes parentheses and brackets to show which operations are to be done in which order. Be on the lookout for new symbols in college math, including certain letters from the Greek alphabet. The notation may be "Greek to you," but it speaks volumes to people who understand it.

Mathematics is not the only discipline that uses its own notation. Music and dance do too. Just as a musician can read sheet music without having to translate the symbols into words, mathematicians read mathematical notation directly and without translation.

Our goal in this chapter is to get you to understand the code once you get out of the habit of translating mathematical statements into words.

First, let's do an experiment. Try "talking" this mathematical equation:

$$\$120/40 = \$3$$

In words, you could state that if 40 identically priced items together cost \$120, then any one of those 40 items must cost \$3. But when you have an expression like a/n where neither a nor n is expressible as a number or something concrete, you have to interpret the expression without using words. Learning to read notation without a word-for-word translation is important because a few pages later you will encounter a mathematical statement that wouldn't be at all easy to translate into words.

$$\frac{(a - V)v}{aTv - RT}$$

Just as an advanced student of a foreign language no longer has to translate word for word, with practice, you will be able to *think* in mathematical notation. That should be your goal.

Understanding Notation

Although the most familiar symbols in mathematics are numbers, letters are used as symbols as or more often than numbers. You met x and y in high school algebra as standing in for "unknowns." Very often one or the other was the answer to a problem you were supposed to solve. They will still be unknowns in college math, but with a meaning: *unspecified*, rather than unknown. They will be called *variables*. Letters at the beginning or the middle of the alphabet tend to represent the opposite of variables: unchanging numerical values. They are called *constants*.

Let's look at an example of how we use variables to express and describe mathematical relationships. Think about the cost of a ride in a cab. Typically, there is a "flag-drop" fee just for picking up a passenger and an additional cost per mile driven. In New York City, a cab driver will charge you a flag-drop fee of $2.50 plus $2.00 for each mile driven. In Chicago, you would pay a $2.25 flag-drop fee and $1.80 for each mile. In Los Angeles, the flag-drop fee is $2.85 and you'll pay $2.70 for each mile.

Let's use variables to describe the relationship between the cost of a cab ride in each city and the number of miles driven. The variable y stands for the total cost of a cab ride and x stands for the number of miles driven. In New York City, we multiply the per-mile rate of $2 by the number of miles, x, then add the flag-drop fee of $2.50:

$$y = 2x + 2.50$$

In Chicago, we multiply the per-mile rate of $1.80 by x and add the flag-drop fee of $2.25:

$$y = 1.80x + 2.25$$

Similarly, in L.A.:

$$y = 2.70x + 2.85$$

What do the equations for all three cities have in common? In each city, there is some unchanging number m (the per-mile rate) and another number b (the flag-drop fee). Using these *constants*, we can write:

$$y = mx + b$$

to express a general formula for the structure of cab fees.

We are typically only interested in the relationship between the cost of the cab ride and the number of miles driven. However, by not specifying the values for the constants m and b, we have a model that works for any cab ride in any city at any time!

One way you can recognize mathematical terms is to know which letters of the English alphabet have special mathematical meanings. Here are a few:

k: typically a constant

i: either an index or the imaginary number $\sqrt{-1}$

e: an irrational number (like π), $e \approx 2.718$

m: slope

f: the name of a function

The code extends to Greek letters, some of which have dedicated meanings. You have most likely seen π for the ratio of the circumference of a circle to the length of its diameter. Others include Σ (*s* in Greek) for summing a series of items and Δ (*d* in Greek) for change in a variable. Often clusters of these symbols are treated as single terms. Examples include Δh, $f(x)$, and $\log(x)$.

Just as in music and dance notation, mathematical notation packs a lot of content. For example, the concept of the logarithm involves four mathematical ideas.[1] All of these ideas are wrapped up neatly when we write $y = \log(x)$.

There is no need to memorize notation. Trust us. The more you see it and the more you use it, the more comfortable you will get. Even though mathematics is not a spoken language, it's helpful to practice speaking new terms out loud so you can "talk mathematics" when asking questions in class. Whenever a new term is introduced, whether in your text or in class, a full explanation will be given. How are you going to remember every new term? Now's the time to think about adding a glossary of notation to your glossary of words!

Subscripts and superscripts are an important part of mathematical notation. You have already learned that the superscript 2 in x^2 signifies x multiplied by itself. The subscript 1 as in x_1 is a way of creating more variables. Remember when there were two students named John in your elementary class? In order to remember who was who, the teacher usually appended the initial of their last names, "John P." and "John Q." This is exactly the same thing that is being done with subscripts. When two variables have the same *first* name (x), we append a subscript, like so: x_1 and x_2, read "x-sub-1" and "x-sub-2."

The key to mastering all of this is to take your time, and get help when you need it from your instructor, your tutor, or your study-mates. Get used to writing or keyboarding symbols. In short, take the new code seriously. We've used the foreign language analogy in this chapter. Mathematics is not technically a foreign language, but you would be wise to study its "vocabulary," its "idioms," and its "grammar" as though it were.

The Equals Sign and Its Relatives

One of the most important symbols in all of mathematics is the equal sign: $=$. You've met the equals sign before. But as you get into higher mathematics, it's important to pay closer attention to it. If an equals sign is found sandwiched between two expressions, it means that the expression on the left side of the equation has *exactly the same value* as the expression on the right side of the equation. For example, if we write:

$$2x + 4 = 8$$

we are communicating that the expression $2x + 4$ has the *same* value as the number 8 in this equation. This degree of equivalence also conveys that one side of the equation can be replaced with the other side.

---

[1] The notions of a function, exponential functions, one-to-one functions, and inverse functions are all wrapped up in the logarithm.

Some students confuse themselves by using the equals symbol to connect steps. This confusion leads to frustration. We have a solution if you feel that you need some way to connect your steps. How about using the symbol $\Rightarrow$?

$$2x + 4 = 8 \Rightarrow 2x = 4$$

Mathematics deals with approximate equality as well as exact equality. There's a difference in signs. The $=$ sign designates exactitude, and the $\approx$ sign designates approximate equivalents. The classic example of an approximate equivalent is the number π. You may remember from middle school math that the value of π is an unending decimal with no repeating pattern. When we use the decimal number 3.14 for π, we are acknowledging that 3.14 is an *approximation* because it would be impossible to write down an *exact* decimal equivalent. To convey this approximation, mathematicians write:[2]

$$\pi \approx 3.14$$

In addition to $=$ and $\approx$, you will encounter some other symbols that describe a relationship between two quantities. Here is a list:

$\neq$: What is on the left is *NOT* the same as what is on the right.

$>$: The value of the expression on the left is LARGER than that on the right.

$<$: The value of the expression on the left is SMALLER than that on the right.

$\geq$: The value on the left is either larger OR equal to that on the right.

$\leq$: The value on the left is either smaller OR equal to that on the right.

The Difference Between "Solve" and "Simplify"

What is the difference between the following assignments?

1. Simplify: $2x + 4 - (x + 1)$.
2. Solve: $2x + 4 = 8$.

The first involves an *expression*. The second involves an *equation*. An expression doesn't tell you very much. Think of it like a noun. If I write "a canine life form" by itself, I haven't really told you much. An equation, on the other hand, is a complete mathematical statement. The equals sign acts as the verb. As you will see, there is a lot more that you can do with an equation than with an expression.

When you are asked to *simplify* an expression, your teacher expects you replace the expression with an equivalent, but somehow simpler, notation. This would be like

---

[2] When students mistake decimal approximations for exact equality, they are frequently mixing up exactness with familiarity. Decimals are comfortable for us to use. They are concrete in the sense that decimal representations extend the place value system that we became comfortable with when learning about whole numbers. The value of 1.41 is clearer to most students than $\sqrt{2}$. There is nothing wrong with this and you have every right to feel this way. What we are suggesting is that you recognize that these values, although very close, are not exactly the same.

changing "the canine life form" to "the dog." In the math problem that we are given, we subtract the expression $x + 1$ from the expression $2x + 4$, resulting in $x + 3$.

Sometimes when comparing algebra problems, it may not be clear which expression is actually simpler. For example, which expression would you prefer to work with, $2x + 4$ or $2(x + 2)$? As you learn more math, you will see that which version of this expression you prefer to work with will depend on some larger problem that you are trying to solve. *Simplify* is one of those terms in math whose meaning is similar to, but not exactly the same as, the meaning in spoken English. What you are really doing when you are simplifying an expression is *rewriting that expression to make it more suitable to a given problem.*

But be careful. When you *simplify an expression,* you cannot change its value. You can only replace an expression with an equivalent expression.

When you are asked to *solve an equation* such as $2x + 4 = 8$, on the other hand, you are being asked to find certain unknown numbers. These numbers are values for the variable x. When you replace the variable in the expression $2x + 4$ with the correct values and compute, you will get the number 8. As we saw, this is what the use of the symbol − demands!

Math instructors commonly use the analogy of scales and balances to differentiate solving and simplifying. Let's say you are asked to simplify an expression. If you think of the expression sitting on one side of a scale, when you replace it with an equivalent expression, the weight *cannot change.* When you are solving an equation, if you think about the two sides of the equation as being balanced on a see-saw, you are allowed to change the weight on each side, *provided the see-saw stays balanced.*

Piecewise Equations

In a higher-level math course, you may come across equations like the following:

$$y = \begin{cases} 2x, & 0 < x \leq 2 \\ 6 + 3(x - 2), & x > 2 \end{cases}$$

Translated into words the meaning is this: If you know x, you compute the value of y with one of two formulas. If x is less than or equal to 2, use the formula $y = 2x$, if x is larger than 2, use the other formula, $y = 6 + 3(x - 2)$. This is called a "piecewise" equation because it has two pieces.

Let's investigate this with a real-world application. Say you go to a parking garage. The garage charges $3 per hour for the first 2 hours, and $2 for each additional hour after that. How can we describe this situation in mathematical terms? We will use the variable x to stand for the number of hours we park, and the variable y to stand for the cost of parking.

There are two possibilities to consider: either we park for less than 2 hours or for more than 2 hours. If we park for less than 2 hours, then $y = 2x$. However, if we park for more than 2 hours, the equation is $y = 6 + 3(x - 2)$. What this means is we pay $6.00 for the first 2 hours and $3 for each additional hour.

The way to convey that our parking cost depends on whether x is smaller or larger than 2 is by using the earlier notation.

These types of equations are called "piecewise" because the first thing you have to do is determine which "piece" of the number line x falls on. Here is a visual way of thinking about this:

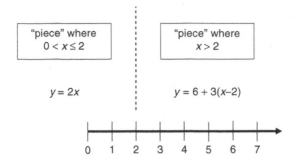

When the number of hours, x, that we park falls in the "piece" to the left of the vertical dotted line (where $x = 2$), we use the equation $y = 2x$ to compute the cost, whereas if x falls in the "piece" to the right, we use the equation $y = 6 + 3(x - 2)$.

Using the Code

Mathematical notation is a powerful code that allows its users to say a great deal in a very precise way without using much space. So in some ways, it's a language; in other ways it's a shorthand. Mastery, and eventually personal ownership, will unlock this code and enable you to apply mathematics to your class work and, as we shall see in Chapters 7 and 8, to your own life as well.

Chapter Four

Algebra Empowered: Functions and Graphs

One sure way to banish math anxiety is to replace it with another emotion. How does substituting anxiety with *power* and *control* sound to you? Well that's exactly where you're going with college algebra. In the algebra you learned in middle school, the letters x and y from the end of the alphabet were used to signify single "unknown" quantities. This quantity you could figure out by following rules pretty much like the rules you learned in arithmetic:

$$3x + 7 = 28$$
$$3x \quad = 28 - 7 \text{ or } 21$$
$$x \quad - 21/3 = 7$$

College-level algebra is different. Here you work with relationships—called *functions*—between two or more unknowns without knowing what numbers the letters in that equation represent. As you move from school mathematics to the real world, you will find that these "functions," mathematical *relationships* between two or more quantities, give you a powerful way of dealing with numerical data in record-keeping, in projecting outcomes, and in calculating the effects of one business decision versus another. These are skills that give the person who knows how to deal with functions an edge.

Variables

When you first encountered variables, they were placeholders for unknown numbers. For example, the x in the equation

$$3 + x = 7$$

represents an unknown number that, when added to 3, yields the number 7. Once you realize that this is another way of asking you to subtract $7 - 3$, you understand why the x stands for the number 4. The variable stood for some *hidden* number, and we can discover this hidden number using arithmetic.

Somewhere in this story, you learned about equations with **two** variables. Although the symbols look the same, the *context* is completely different.

Let's start with an equation that looks like it can be reduced to a single "answer" like the previous problem, but really can't because there are two unknowns.

$$y = \frac{1}{2}x$$

It is not possible to solve this equation without more information because although the two unknowns have a relationship, there are (as it turns out) lots of y's that could equal $\frac{1}{2}x$ depending on the value of x. Once we set the value of x, the value of y is completely determined. The value of y depends on the value assigned to x. For example:

$$
\begin{array}{ll}
\text{when } x = 2 & y \text{ has to} = 1 \\
\text{when } x = 4 & y \text{ has to} = 2 \\
\text{when } x = 4.5 & y \text{ has to} = 2.25
\end{array}
$$

Aha! There is something unchanging here. Not the numbers, but the *relationship*. The unknown y is always equal to the unknown x divided by 2. This apparently simple algebraic equivalence is more than a statement about two unknowns leaving us without knowing the value of either. It is a description of an *unchanging* relationship between the two values.

Suppose you want to calculate how much gasoline you could purchase for $16 that you have on hand. One gas station is offering gas at $2 per gallon. At this gas station, you would divide your $16 by $2 per gallon, which would tell you that you can purchase 8 gallons of gas with the money you have in your pocket.

Suppose next week, you have $9.50 to spend on gas. You do the same calculation again.

Here is where we can use the power of algebra (in two variables) to allow us to see how many gallons of gas you can buy with any sum of money at this gas station:

$$y = \frac{1}{2}x$$

The power comes when you graph this equation and can look at the relationship. You will *see* all possible solutions at once. This example may seem silly—why don't we simply divide by 2 each time we need gas? But what if the gas station also charged a $5 fee for using a debit card, and there is a gas station across the street that charges $3 per gallon and no fee? The power of algebra and graphs will allow us to decide on a simple rule that will tell us which gas station we should go to without having to separately compute which is cheaper in every possible situation.

Unlike the original use of the letter x in the equation $3 + x = 7$, we use *variables* in this context not because we do not know some numbers, but because we *choose* not to specify their values. Although we certainly could choose how much money we want to spend on gas and just divide that number by 2, by writing an equation relating the two variables, we can *predict* how much it will cost for any possible amount of money. Algebra also empowers us to *generalize*—to describe multiple situations (such as the cost of gas at different gas stations). In this way, we can use our predictions to make decisions.

Functions

Mathematicians have special words for relationships between quantities (such as the amount of money available for gas and the amount of gas that can be purchased). Mathematicians call these relationships **functions**. The variable x is the **independent variable**, and y is the **dependent variable**. We use the words *independent* and *dependent* because we can choose the value of x independently (such as $x = 16$), and the value of y will depend on this choice ($y = 8$).

Now why can't we do the reverse? Mathematicians have an agreement (they use the word *convention*) that the letter x will always stand for the independent variable and y will stand for the dependent variable. This way, everyone speaks the same language.

On the other hand, relationships *can* be reversed. For example, for the gas station charging \$2 per gallon, we have been talking about how much gas we can buy with x dollars. However, we can also talk about how much it costs to buy x gallons of gas. In this case, $y = 2x$ (where y is the cost of purchasing x gallons of gas). We are using the same *symbols* as variables, but their meaning has changed and their relationship has switched.

Graphs: The Power of Graphing Equations

What is new and powerful about x and y as variables is that we can "plot" (that is, draw) their relationship on a graph. Graphs allow us to look at a function such as $y = \frac{1}{2}x$ and see the *relationship* between the two values, y and x. Instead of having to perform a computation every time we spend x dollars on gas, we can see, visually, the effect on y of changing x. So it is really important to know how to construct and read graphs.

In order to graph the relationship described by a function, we should understand what graphs look like and how to read them. Look at the following map of **Ohio**:

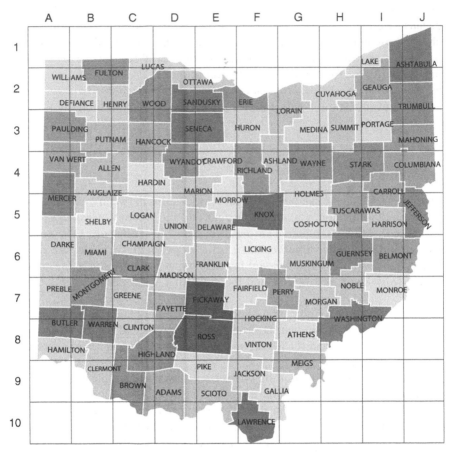

According to the index, **Montgomery County** is located at B-7 on the map. To find this location, we look for the column labeled B. We see our location where the row labeled 7 intersects the column labeled B.

Graphs are used everywhere and for lots of purposes. For example, the following graph shows the price of a share of stock in **Tobias Motors** over the course of February 23, 2012.

Now on *this* graph, the horizontal axis tells us the time, and the vertical axis the price. If we look at the column corresponding to 1:00 p.m., we look at the point on the broken line and see that its corresponding row is $26.78. This means that at 1:00 p.m., 1 share of the company's stock was selling for $26.78.

Sometimes we see graphs in different forms. For example, the bar graph on page 35 shows the unemployment rate (as a percent) in each month of 2011. In

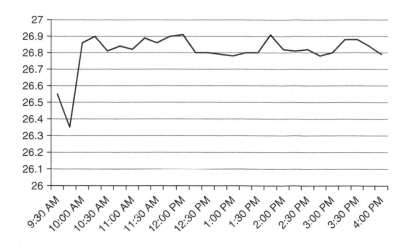

a typical textbook problem, you might be asked to find the unemployment rate for July 2011, or to find how much the unemployment rate changed between July 2011 and December 2011.

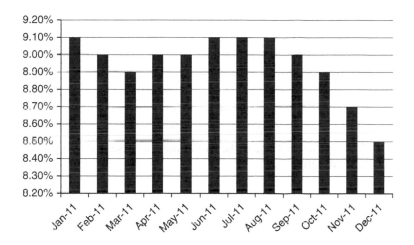

When we look at the bar for July 2011, we see that the unemployment rate was 9.1 percent. In December, the rate was 8.5 percent. Now we can find the change by simply subtracting!

All of these examples show that graphs give us the power to see a relationship between two quantities, and we can see all of the information at once! For example, look at the Tobias Motors graph. When was the stock price increasing? When was the stock price decreasing? When would be the best times to buy and sell this stock during the day? The graph allows us to take data found in the news or in nature and answer these kinds of questions.

The Cartesian Plane

Here is what we mean by algebra empowered: the connection between a function and graphs. In order to construct the graph of a function such as $y = \frac{1}{2}x$, we need to construct a **Cartesian plane**.[1]

Much like the graphs we just read, the *horizontal axis* allows us to track the independent variable. This is the variable that we have control over. It looks just like the real number line.

Next we add a *vertical axis*, which allows us to track the dependent variable. This looks like another number line, this time pointing vertically, with numbers increasing as we move up the graph. Together, the x and y axes create a *grid*:

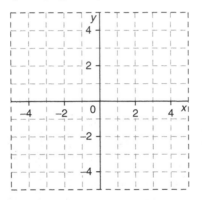

Each point in the plane, just like in the earlier map, corresponds to a pair of numbers, called **coordinates**.

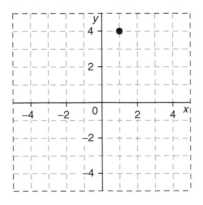

---

[1] This coordinate system was created by and named after the philosopher and scientist Rene Descartes in 1637.

To find the coordinates of this point, we drop a line straight down to see what the x-coordinate is, and a line straight across to find its y-coordinate.

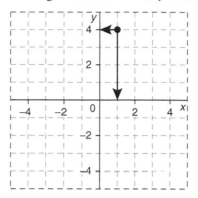

The coordinates of this point are $(1, 4)$. This is known as an **ordered pair**. Another convention that mathematicians have agreed to over the years is that the value of the independent variable always comes *first* in an ordered pair.

Given an ordered pair, such as $(-1, 3)$, we plot the point just as we read a map. We find the column for -1, and the row for 3. Where the column and row intersect, we plot our point.

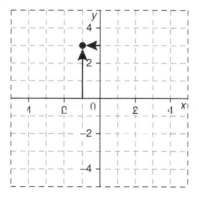

Compare the point $(-1, 3)$ that we just plotted to the point $(3, -1)$:

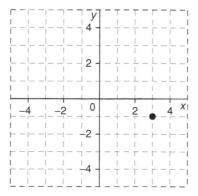

These are not the same point. The reason we are showing this to you is to empha-
size that *order matters*. That's why we use the term *ordered* pairs.

Graphing Functions

The world around us speaks to us with data, as we saw in the graphs. Graphs
allow us to process that data, and if we can find functions whose graphs fit the
data, we can answer several questions. This process is called **modeling.** Although there are some schools that are moving in the direction of starting with
data and graphs leading to algebraic functions, most schools are still starting
with a function and asking you to construct the graph and answer questions.
The important point here is the correspondence between the algebra in the
function and its graph. Being able to move fluidly between one and the other
in any direction will empower you to answer what would otherwise be difficult
questions!

Let us return to our equation $y = \dfrac{1}{2}x$. To construct the graph, we need to
write down some of the many possible solutions to this equation. These solutions
consist of two numbers, a value of each of the variables, which when replaced
for the variables makes the statement true. At the beginning of this chapter, we
found a few solutions to this equation:

$$x = 2, y = 1;$$
$$x = 4, y = 2;$$
$$x = 9, y = 4.5.$$

These are usually written as ordered pairs: (2, 1), (4, 2), and (9, 4.5).
Now we plot the ordered pairs on a graph:

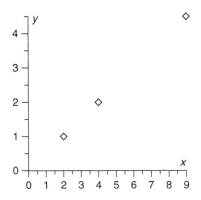

Graphs of functions that have the form $y = mx + b$ are straight lines. The values of
m and b are usually given to you as numbers. In our example, m is ½ and b is 0.

The graph of this function is a straight line that must pass through the points that we plotted:

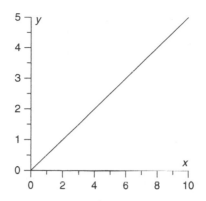

We now can use the graph to understand the relationship between the number of gallons we purchase and the price we pay. The graph represents all ordered pairs that are solutions to the equation $y = \frac{1}{2}x$. Although there are infinitely many solutions, the graph, in a sense, allows us to *see* all of them at once! This allows us to explore the relationship.

Before we address questions that we can answer with this graph, you might be wondering how we knew that this graph would be a straight line. It turns out that all functions of the form $y = mx + b$ have graphs that are straight lines (here m and b are supposed to be fixed numbers, for example $m = \frac{1}{2}$ and $b = 0$). Such equations are called **linear**, precisely because their graphs are lines! This is what you should be looking for out of your math course—what types of algebraic equations correspond to what shapes for graphs?

For functions whose equations have the form $y = mx + b$, the number represented by m is probably the most important. This is the *slope* of the line. The larger this number, the steeper the line.

Using the Graph: YOU Have the Power!

Let us now use the graph of the equation $y = \frac{1}{2}x$ to answer some questions about gas prices.

Question 1: How much gas can I purchase with $10.50?

Suppose I have $10.50 to spend on gas. How much can I afford? We read the graph like a map. We find the value 10.5 on the vertical axis, which represents gallons. Draw a vertical line across the grid through this point on the horizontal axis. The y-coordinate of the intersection point with the graph tells us how much gas we can afford!

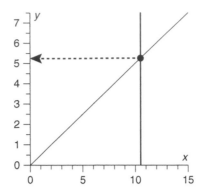

This tells us that we can afford approximately 5.25 gallons of gas with $10.50.

Note that the algebraic definition of the function, $y = \dfrac{1}{2}x$, tells us that we figure this out by cutting the number of dollars that we have in half. Although this may be simple enough, one can imagine far more complicated functions (examples of which we will point out later in the chapter) for which a graphical estimation may be much more manageable.

Question 2: How much does it cost to buy 10 gallons of gas?

Imagine that it will take 10 gallons of gas to fill your tank. How much will this cost? We invert the process: find 10 on the vertical axis, which represents gallons of gas. Draw a horizontal line across the grid through this point.

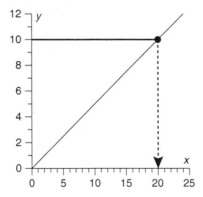

The x-coordinate of the intersection point with the graph tells us how much this will cost—in this case, $20.

What is with the horizontal and vertical lines? What is really going on here?

All of this drawing of horizontal lines and vertical lines may seem confusing. What is really going on here is that we are taking a two-variable equation and, when we know the value for one variable, we can find the other using algebra.

The graph is allowing us to do the algebra problem visually! For example, in the second question, you are really solving the one-variable equation:

$$\frac{1}{2}x = 10$$

In this case, you know the y-value but not the x-value. The point where the horizontal line through $y = 10$ intersects the graph of the function tells us the solution—we simply read off the x-coordinates. For some functions, there may be more than one!

In school, you are ordinarily introduced to equations with one unknown first. You learn how to find the solution to an equation such as:

$$\frac{1}{2}x = 10$$

Once you have mastered equations with one unknown, you are introduced to equations with two variables. We think it is easier to *begin* with two variables. Then you can use the *graph* to actually *see* equations with one variable and their solutions.

Comparing Alternative Situations

Imagine that one gas station sells gas at $2 per gallon and another sells gas at $3 per gallon. How do we know when to go to one gas station and when to go to another? Although we could compare the graphs $y = \frac{1}{2}x$ and $y = \frac{1}{3}x$, it might seem silly because it is obvious that you should go to the gas station with the lower price. If we actually did plot these graphs, we would see:

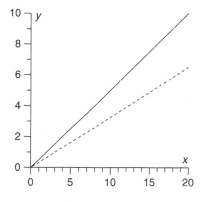

The solid line is the graph for $y = \frac{1}{2}x$ and the dotted line is the graph for $y - \frac{1}{3}x$. The graph for $y = \frac{1}{2}x$ always lies above the graph for $y = \frac{1}{3}x$. This means that for any fixed amount of money (x), you will be able to afford more at the first gas station. So here, the graph confirms our common sense.

Now imagine that the first gas station that charges \$2 per gallon also charges a \$5 fee for using your debit card (you do not have any cash with you). Which gas station would you use then? Let us model and graph!

In this case, it is easier to think about this if we reverse the relationships. If you purchase x gallons of gas at this gas station, you pay $2x$ dollars PLUS a fixed \$5 fee. Your new function is thus: $y = 2x + 5$. The graph is as follows:

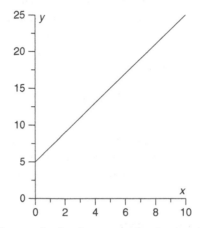

Let us now compare the graphs for the two stations.

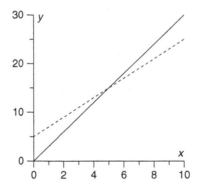

The solid line is the graph of $y = 3x$ and the dotted line is the graph of $y = 2x + 5$. In this graph, you can see that to the left of the intersection point of the graphs, it is cheaper to actually use the higher gas price at the second station. To the right, if you are purchasing more gallons of gas than at the intersection, it is cheaper to go to the first station and suck up the fee.

How do we find this intersection precisely? We can use the graph, but if there is any uncertainty, we can use algebra. At the intersection point, the co-ordinates of the point on both graphs must be the same. This doesn't help if we only think of the x-coordinate, but it does with the y-coordinate. Using the relationship, finding where the intersection occurs amounts to solving the equation with one variable:

$$3x = 2x + 5$$

Solving this equation using algebra, we obtain the solution $x = 5$. Remember that x is measured in gallons. This means that if you are purchasing less than 5 gallons of gas, go to station 2, whereas if you are purchasing more than 5 gallons of gas, go to station 1. If you are purchasing exactly 5 gallons, it doesn't matter which station you choose.

We find ourselves going back and forth and forth and back between the algebra and the graph. If we want to figure out a rule allowing us to decide which gas station to go to, we can find equations (algebra) for the cost of x gallons of gas. These equations don't tell us much, but if we plot their graphs, we *see* what is going on—the picture tells us how to make the decision. But to find the exact point where the graphs intersect—the threshold between choosing one gas station and the next—we have to go *back* to the equations and do some algebra. In this problem, the *graph* tells us what *algebra* problem we have to solve! But we don't have a graph unless we do some algebra first.

The Art of Noticing

Look at the graph of the equation $y = \frac{1}{2}x$.

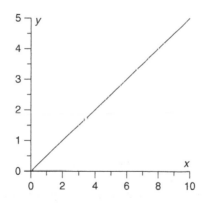

What do you notice about the graph? This question may make you feel uncomfortable. Your first reaction might be to ask "what am I supposed to notice?" If you ask yourself this question, it might be because you believe that you are supposed to come up with right or wrong answers. By telling yourself that, you are limiting your ability to discover and own your observations. This limits your power. Give yourself permission to explore the graph, free of worry that what you say may be "wrong" or "useless."

Suppose you say "the graph points upward, what does this mean?" This is actually important. If you ask your instructor, she might respond by saying "It means that the function is increasing—when you make x bigger, y gets bigger also." It may seem fairly obvious that the more money you spend on gas, the more gas you can afford. That the graph agrees with our common sense is a signal that our model is *reasonable*.

Now look at some other graphs and see what you notice:

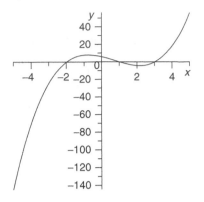

What we notice here is that this graph has some bumps, called *local maxima* and *local minima*. We also notice that the graph crosses the *x*-axis three times.

Try another graph:

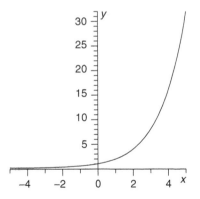

The first thing we notice is that the graph is always pointed upward as we move along the *x*-axis from left to right. In mathematical terms, this function is *increasing*. We also notice that it looks as though the graph never crosses the *x*-axis.

Try to do the same thing with the graph on page 45 showing the price for a share of Tobias Motors stock over the course of February 23, 2012.

Imagine that you are looking to make as much money as you can by buying and selling 1,000 shares of this stock during the day. In your list of things that you noticed, what do you think would be important in making this decision?

Looking at a picture and "noticing" things is an art, not a science. There aren't any right or wrong answers. Think of this as a chance to explore—explore a picture and see what you can find. This is something that you can do not just by yourself, but with a study group. What you notice that may or may not be useful will depend upon the context of the problem. In your math class, you will learn what is important to mathematicians.

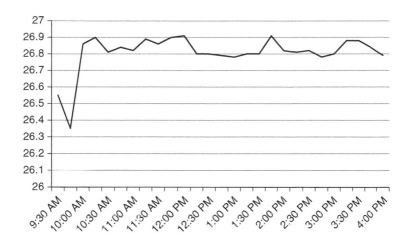

Conclusion: Look for the Power!

You should think about your math course as learning the skills that you need to probe a function. The graph allows you to see everything at once; the equation allows you to manipulate and find precise answers to the questions you will learn to answer. Students who are successful in math find themselves going back and forth and forth and back between the graph and the equation. Even when a problem seems purely algebraic, graphing allows you to visualize what you are computing. This is the power of algebra—the ability to convert back and forth between equations and graphs in order to make predictions and make decisions.

Short Take 3

Practicing Function Notation

Back in Chapter 3, we showed you how to decipher notation: how to translate math symbols. As we saw, notation is a shorthand. For functions, if we have an equation such as:

$$y = 2x + 5$$

mathematicians will write $f(x) = 2x + 5$. Here, the clump of symbols $f(x)$ is shorthand for the *function* described by this equation. Then what is f? Why the parentheses?

Mathematicians use the symbol f to refer to the function itself—the relationship between x and y. The parentheses act like a machine. We put a number, say 6, into the parentheses (the "machine") and some other number pops out! For this function, the machine multiplies the number 6 by 2 and adds 5, popping out 17. The shorthand mathematicians write is $f(6) = 17$.

Why don't mathematicians just say that? Well, this can be quite a mouthful! By writing $f(6) = 17$, the mathematician is saying that when $x = 6$ is put into the function f, the output is 17. The mathematician also uses this shorthand to remind us that there is a relationship between x and y.

The parentheses may be the most confusing part of the notation. In pre-algebra, you may have been taught that parentheses mean "multiply." If you see the quantity $3(x + 5)$, you write $3x + 15$. In this context the parentheses are a grouping symbol—they tell you which computation to do first.

Now the parentheses are being used in an entirely new way. This is the same as when a word has more than one meaning. Think about the word *pet*. This word has one meaning in the sentence "I have a *pet*" and an entirely different meaning in the sentence "I will *pet* the dog." How do you know which meaning to use? The context of the sentence makes it clear.

So what do the parentheses mean in function notation? The parentheses are a *place*. The variable x is a placeholder. The placeholder could be replaced with a number or even another expression, depending on what you are being asked to do. If I am using the function $f(x) = 2x + 5$, and I write $f(3)$, this means I replace x with 3 and simplify:

$$f(3) = 2 \times 3 + 5 = 11.$$

With the same function, I can write:

$$f(x + 4) = 2(x + 4) + 5$$
$$f(a + h) = 2(a + h) + 5$$

What you are being asked to do in these types of problems is to demonstrate your familiarity with substitution and your comfort level with this notation. Function notation will be extremely useful if you ever take advanced math courses such as calculus.

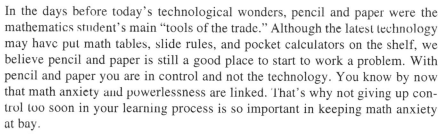

Chapter Five

Tools of the Trade

In the days before today's technological wonders, pencil and paper were the mathematics student's main "tools of the trade." Although the latest technology may have put math tables, slide rules, and pocket calculators on the shelf, we believe pencil and paper is still a good place to start to work a problem. With pencil and paper you are in control and not the technology. You know by now that math anxiety and powerlessness are linked. That's why not giving up control too soon in your learning process is so important in keeping math anxiety at bay.

It's no accident that mathematicians prefer pencils to pens. With a pencil comes an eraser, which gives the user permission to make mistakes and correct them. You can take a cue from the professionals. You can learn a lot from your mistakes. They can be windows into your thinking.

If you watch professional mathematicians do their own work (not when they are teaching you how to do yours), chances are they will be writing by hand on a chalkboard, if not in a notebook or on a pad.

That's not because they don't know how to use advanced calculators and mathematical software. They can and they do. But before they light on a solution to a problem, they need to have a plan. They feel they can work out an approach better by hand, given enough time.

Our goal in this chapter is to make you feel comfortable with using paper and pencil at a certain level of problem solving in math, after which we will guide you in the use of more modern mathematical tools.

Paper-and-Pencil Explorations

Let's explore some paper-and-pencil mathematics. Because, as we have seen in the last chapter, plotting a graph on graph paper is a means to understanding functions, we'll begin by doing some plotting by hand. Don't forget your eraser!

We'll start with a classic: $y = x^2$.

To start plotting this equation, we choose some x's and compute the corresponding y's. This is done by constructing a table of values for x and computing the corresponding values for y.

Constructing tables of values

How do we know what range of values of x to choose? If you know something about the function before you plot its graph, that will help. It's often a good idea to include both positive *and* negative whole numbers. If we can't see enough of the graph, we can always plot more points.

We usually write our chosen x-values in a chart:

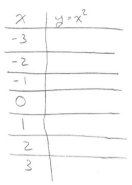

The numbers don't have to be organized in any particular order, but organization will help you see patterns.

Once we have the x values down, we can square each x to find the corresponding values of y.

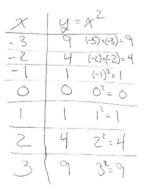

Plotting the function

The next step is to mark the points on graph paper. Compare Graph 1 and Graph 2. Graph 1 shows the $y = x^2$ very clearly. On Graph 2, the points we're trying to represent on the graph are scrunched together. This has to do with scaling.

Graph 1   **Graph 2**

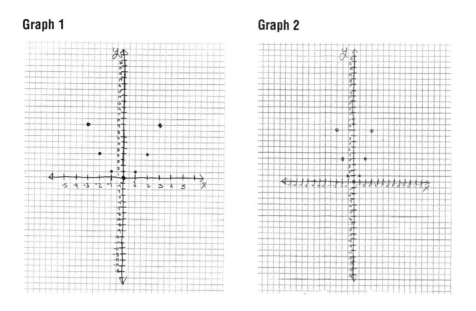

After plotting points, we connect the dots, sketching from left to right.

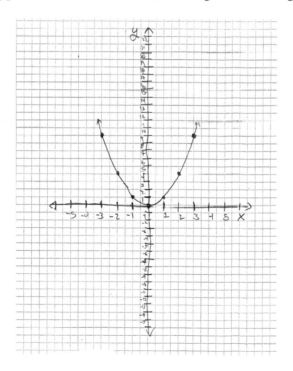

As you get more experience in marking points on a graph and connecting them, you will begin to recognize why visualizing the behavior of a function is such a powerful means of understanding the function. Look again at the last graph. Clearly, it doesn't pay to continue to add more points to the graph. One can see by looking at it that the curve is going to continue in the same manner no matter how many more points you would plot.[1] This is not the case with the next example.

Another example

Let's try to plot the points on graph paper for the function $y = x^3 - 12x$. When we write x^3, this means x times x times x. We will try the same x-values that we used in the last example:

Next we plot the points and connect them:

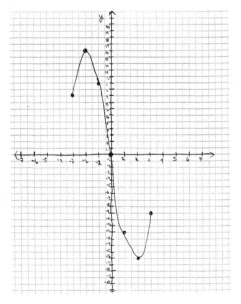

_____

[1] The experienced math student—eventually, after plotting lots of simple equations on graph paper—will actually be able to guess at the shape of the curve.

Are we finished? This isn't as good of an illustration as the last example—we don't know where the graph goes after -3 and $+3$. Why don't we try some more points?

If we substitute $x = -4$ into the function, we find $y = -16$. If we substitute $x = 4$, we find $y = 16$. We add the points to our plot and connect them:

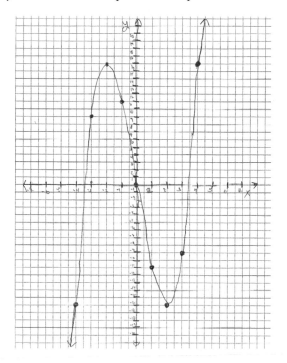

Aha! We can see more of the behavior of the function. We can even see that in addition the point where $x = 0$, the graph crosses the x-axis between $x = -4$ and $x = -3$ as well as at some point between $x = 3$ and $x = 4$. If you were asked to find the exact x-values for these *intercepts*, you would go back to algebra.

In this last example, the hand computations involved more steps than simply squaring. Other types of functions will require still more computations that you will not want to calculate by hand. At this point you will want to be able to use a scientific or graphing calculator.

The Road to Graphing Calculators

Before the first scientific calculator came on the market, a tool called a *slide rule* was used to multiply and divide very large numbers. The first electronic calculators were designed to handle arithmetic computations. They couldn't do much else, but they were greatly appreciated for the time and effort they saved. In 1971 Hewlett Packard developed the first scientific calculator to handle higher-level functions beyond arithmetic.

Today, you don't have to buy a scientific calculator. You can download it onto your smartphone.

The calculator that you can expect to use in your math class is the *hand-held graphing calculator*. It was developed in the 1980s. Today, graphing calculators are used mainly for classroom instruction. Even if the class you are taking doesn't involve a graphing calculator, one of your future math courses is likely to require its use.

When it comes to graphing functions on a graphing calculator, one thing that's new and different from plotting by hand is the graphing calculator's "window" menu. The window specifies how much of a graph you want to be able to view.

Here is one student's graph of a particular function:

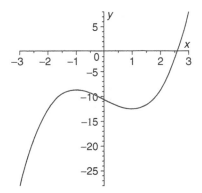

This window is a good choice because it allows us to see the main features of the function, such as where the curve crosses the *x*- and *y*-axes and its local minimum and maximum.

If the student had selected another window, say the following window, the resulting graph might be less informative:

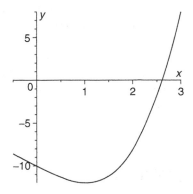

This particular window affects what you see.

Choosing a window for your graphing calculator is similar to selecting which *x*-values you are going to use to plot a function by hand. If you already know something about the function, that will help. If not, you can always just try something, no matter how arbitrary. Then you can use what you are learning in your math class to decide if the graph seems reasonable.

"Calculator anxiety"

When you start using a graphing calculator, you could fall into one of two traps: (1) overreliance on the calculator, forgetting that you need to get comfortable with the new content by doing problems by hand; or (2) being overwhelmed by the calculator's rigidity and complexity. Too much reliance on the calculator—the first case—will leave you feeling helpless when you're not allowed to use it on a test or when the battery runs dead. Try doing your graphing calculator assignments both ways: by hand and with the help of the calculator.

As you get more comfortable with your calculator, you will start noticing patterns in the commands. These patterns will guide you to the calculator's own language, commonly called its *syntax*. Once you're comfortable with the calculator's syntax, new commands will make more sense.

Beyond graphing calculators

Today, professionals use computer software packages specific to their fields in addition to calculators. The most common software packages, which even do calculus symbolically, are Matlab, Mathematica, and Maple.[2] To accommodate their needs, computer scientists have created specific software for solving problems in statistics, physics, chemistry, engineering, finance, and other specialized disciplines that involve math.

Your graphing calculator, then, is worth mastering as the first of a vast array of programming tools that you may have to learn depending on what you end up doing professionally.

Online Tools: Wikipedia, WolframAlpha, and Khan Academy

Several Internet resources can be helpful if used properly. Professionals make extensive use of Wikipedia and WolframAlpha. Students find the explanations given at Khan Academy to be very clear. We encourage you to follow along on the Internet as you read.

Online tools used by professionals

Wikipedia, an online encyclopedia, is a widely used resource. If you enter the word "function" into Wikipedia, you get a list of several pages corresponding to that word. You are interested in the mathematical notion, so you click on the link for "function (Mathematics)." On the page for this entry, there is an introduction, followed by a table of contents, followed by the contents themselves. The table of contents is clickable, so if you are particularly interested in something specific, you can click on the link that will take you directly to that entry.

In addition, throughout the page, you will find links for terms that you can click on. If you do not understand a term that is used on the page, for example, the term *real numbers*, you can click on the link. Once you are done, you can browse back to the original page. Of course, on the page for "real numbers," there may be

---

[2] As an example of how useful these software packages are, all of the computer-generated graphs in the book were created with Maple.

a term you don't understand that you can jump to as well. This is one of the great advantages of Wikipedia. It allows you to jump between pages so that you can get a complete understanding of the topic you are investigating.

There are disadvantages to using Wikipedia. One disadvantage is that the descriptions may be too abstract or theoretical for your purposes. If this is the case, ignore it. Another disadvantage is that anybody can alter the content, which limits the site's reliability overall.

A newer resource used by professionals is starting to catch on in college classes. This is WolframAlpha, found at www.wolframalpha.com.

From www.Wolframalpha.com by WolframAlpha. Copyright © 2012 by WolframAlpha. Reprinted by permission.

This website calls itself a "computational knowledge engine." One advantage of WolframAlpha over Wikipedia is its reliability. In addition, it displays results in a very concise manner and is unlikely to have paragraphs with abstract definitions. However, several results may be too advanced for your purposes.

Suppose you see a reference to the "sine function." This isn't something you've heard of before, so you decide to investigate. You type "sin(x)" into the search bar on WolframAlpha, and a wealth of information about the sine function is displayed. We see, for example, two graphs in different windows.

From www.Wolframalpha.com by WolframAlpha. Copyright © 2012 by WolframAlpha. Reprinted by permission.

If we scroll further down, we find the roots, that is, the solutions to the equation $\sin(x) = 0$. There is also information that is more advanced than you need. For example, the program displays information about derivatives and integrals, which belong in a calculus course. All that information can be quite overwhelming!

Online resources for professionals should be used selectively. Do not let the wealth of information overwhelm you. Much of what you find is meant for students in higher-level classes. By writing down specific questions before beginning your work, you will find it easier to obtain the information you seek.

Khan academy

Khan Academy is a web page that contains short YouTube lectures on individual topics for many courses—and not just math! Most math videos are 10 to 15 minutes long and contain several clear examples. When you watch a math video on Khan Academy, you won't see someone standing at a board lecturing to you. Instead, you will see the board itself. As you listen to the speaker, "markers" will be writing on the board in multiple colors. Many students say they find the explanations to be very clear.

One advantage of using online videos such as those from Khan Academy is that you can control the pace. In class, if you didn't catch something the instructor said, or your instructor erased something on the board before you were done with it, your only option is to ask the instructor to repeat or rewrite. With Khan Academy on your screen, you can rewind the video or move it forward as you please.

Mastering the Tools of the Trade

However you're learning new mathematical skills, learning how to use the tools and resources available to you puts you back in the driver's seat. That's true of calculators, online instructional tools, textbooks, manuals, software packages, online help sessions, and paper and pencil.

Short Take 4

Playing with Your Graphing Calculator

Think of your calculator as a laboratory where you can conduct experiments, try new things, and really *play*. Just like explorers need tools to travel and discover the wilderness, you can use your graphing calculator to *explore* the landscape of math! This could be a fun way to spend time with your study group. In this short take, we walk you through a couple of explorations you can conduct.

Exploration 1: What causes some lines to be steeper than others?

Graph these two equations simultaneously on your calculator:

$$y = 4x + 3$$
$$y = 2x + 3$$

What do you notice? Which line is steeper?
Here is what we see:

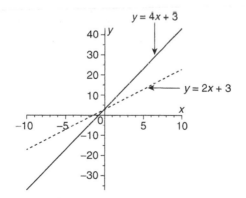

The graph of the function $y = 4x + 3$ is a steeper line than the graph of the function $y = 2x + 3$. This is because the first function has a larger *slope*—the number that is multiplied by x in the equation. The first slope is 4, and the second slope is 2. The larger the slope, the steeper the line.

What about negative slopes? Try graphing and comparing the functions $y = 2x + 3$ and $y = -4x + 3$.

Here is what we see:

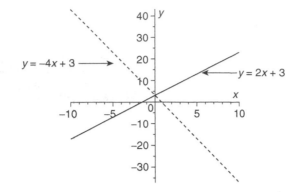

What we notice here is that the line with the positive slope points "upward" from left to right; it is *increasing*. The line with the negative slope points "downward" from left to right; it is *decreasing*. This is exactly the difference between a line with a positive slope and a line with a negative slope.

Exploration 2: What happens if you try to divide 1 by 0?

In addition to graphing functions, graphing calculators can produce tables of values. We will use both of these features to investigate this question.

First, type $1 \div 0$ into your calculator. What happens? You will get some error message, such as "ERR: DIVIDE BY 0." This might not surprise you if you remember being told over and over again that you cannot divide by zero.

How can we explore what is actually happening *near* 0?

Let's look at another graph. We can use algebra to write a formula for dividing 1 by some number by replacing "some number" with—you guessed it—a variable! So we divide 1 by x, and say that the quotient is the variable y. Using mathematical notation to write an equation:

$$y = \frac{1}{x}$$

Here is the graph of this function:

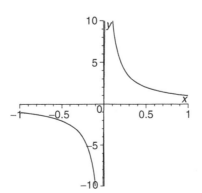

On the left side, the graph plunges downward. On the right side, the graph flies upward.

In addition to looking at the graph, we can explore *numerically*—we can look at the actual numbers that you get when you divide 1 by specific values of x. Using the table tool, the first table that was reported to us was:

X	Y1
1000	.001
2000	5E-4
3000	3.3E-4
4000	2.4E-4
5000	2E-4
6000	1.7F-4
7000	1.4E-4

Just like using the window in a graphing calculator to control what you are looking at, you can also tell the calculator what values you want in the table. For instance, we had the calculator create the following table:

X	Y1
−1	−1
.1	10
−.01	−100
1	1
.1	10
.01	100
.001	1000

In this table, we notice that as *positive* numbers get closer to zero, the quotients get very large. As *negative* numbers get closer to zero, the quotients, in contrast, get very large in absolute value, but are negative.

Try entering values that are getting closer and closer to zero. Use both positive *and* negative numbers. Is our observation still valid? Why is this happening? How is this reflected in the graph? These are questions that you can talk about in your study group and with your instructor.

Chapter Six

Problem Solving

Is there one "scientific" way to solve a problem in mathematics? It would be reassuring to think there is a step-by-step procedure that works every time. But there is no one guaranteed "right" way to solve all problems. Some methods may be better than others. One solution may be more efficient or, as mathematicians put it, more "elegant"; another may be more useful in the long run because the same technique can be applied to more complex examples. But problem solving is also very individual. What works for one problem solver may not work for another. What works in one type of problem may not work in another. In studying math, you probably have come across theories about the four, six, or ten "steps" in problem solving, organized in some strict sequence. But almost no one ever *discovers* solutions the first time working so rigidly. Experienced problem solvers first look for a strategy, *any* strategy, that will lead to a solution. After you have arrived at a solution, it is easy to see how you *ought* to have gone about finding it. But few people start out as systematically as they would like.

Strategies

Here's something your math teacher may not have told you: successful students of math use a variety of problem-solving strategies, but—and this is important—not in any rigid order. They start by exploring the problem. What does the problem tell me? What does the problem ask me to find out? Have I seen similar problems before? Do I need to use a formula, such as the quadratic formula? Then and only then they begin to work.

You have an initial *strategy* in each case. You work in a somewhat haphazard way. You try to think of the rule that applies in each particular case. And then you apply the rule. When you get your first answer, which may not be your last answer, you *check* your answer. You monitor your progress. Sometimes you have to shift your strategy when you notice some detail that you overlooked when you started out. You use several different tasks throughout the process:

- Initial strategy
- Shifting strategies
- Searching
- Sorting

- Guess and check
- Refining the guesses
- Checking how you are doing
- Continuing to search

These are not steps but strategies you will use in tandem, going from one to the other and back again, experimenting and improvising as you reach for a solution.

A Typical Problem: When Does Solar Power Pay for Itself?

Let's take a look at a fairly typical textbook problem involving solar power.

Solar power involves harnessing the energy of the sun to power a home. Certain panels that collect solar energy must be installed, usually on the roof of the home. In order to hold up the additional weight, the roof often needs to be strengthened and pipes may have to be rerouted. Finally, the solar panels must be connected to a generator, which in turn must be connected to the electrical system of the house. With all this work, a solar power system can be quite costly to install. However, after time, a homeowner can save a considerable amount of money by converting to solar electricity. At what point does the homeowner begin to save money by this arrangement?

The following problem, and the numbers that are involved, are adapted from a 2003 textbook.[1] By the time you are reading this, the costs for solar power will most likely be lower.

Suppose that for a three-bedroom house in your favorite community, the cost to install a solar power system is $30,000 and the cost to install an electrical system connected to the community power lines is $5,000. The maintenance cost for the solar power is $100 per year, but you do not have to pay a utility company for your electricity. The traditional electrical system has minimal maintenance costs, but requires the owner to pay $1,100 per year to the electric company.

Problem: *When does the solar power system start to pay for itself? In other words, after how many years will the total cost of installing the solar power system balance out the savings from not paying the monthly electric bill?*

Let's look at the way the cost of both systems is calculated. First, we have the cost of the installation. Then we have two different annual costs to compute: maintenance for the solar power system and electric bills for the traditional system.

Your goal is to solve the problem. But to do this, you have to start by writing an equation. In ninth grade you would just be given the equation. In college math, you will be given the problem and asked to create the equation. Here's how to do this.

First, an experienced problem solver will "play" with some numbers to get a sense of how the costs of the two systems are computed over certain time periods. This is the exploratory part of solving this problem.

---

[1] Robert Blitzer, *College Algebra,* Prentice Custom Publishing (for Michigan State University), 2003, p. 138.

Here's how the costs of the two systems are computed after 5 years:

System	Installation	Annual Cost	Total Cost
Solar	$30,000	$100 × 5 = $500	$30,500
Traditional	$5,000	$1,100 × 5 = $5,500	$10,500

Our conclusion, as would be the homeowner's, is that after the first 5 years, it is cheaper to have stayed with the traditional electrical system.

What about after 30 years? Assuming the annual cost doesn't change:

System	Installation	Annual Cost	Total Cost
Solar	$30,000	$100 × 30 = $3,000	$33,000
Traditional	$5,000	$1,100 × 30 = $33,000	$38,000

Now the total cost of the solar power system is cheaper! At what point does the cost shift over? We know from what we've calculated so far that it's sometime between 5 years and 30 years. Do we have to keep making new tables? Not if we can find a pattern that will enable us to create an equation.

In order to find a pattern, let us look at how we computed the total cost for each year.

For the **solar power** system, look at the pattern in the computations displayed in the table:

Year	Installation	Annual Cost	Total Cost
5	$30,000	$100 × 50 = $500	$30,500
30	$30,000	$100 × 30 = $3,000	$33,000

What's going on here? What is changing? What has varied? The two quantities that are changing are the number of years and the total cost. These are the quantities that must be represented by variables. It is helpful to let our independent variable, the x, be our unknown answer. So we will assign the variable x to the number of years and the variable y will be the total cost. To obtain the total cost y, we add $100x$ to $30,000. Our equation is:

$$y = 100x + 30,000$$

Let's look at the traditional, **non-solar** electrical system the same way:

Year	Installation	Annual Cost	Total Cost
5	$5,000	$1,100 × 5 = $5,500	$10,500
30	$5,000	$1,100 × 30 = $33,000	$38,000

Here we multiplied the annual cost paid to the electric company by the number of years and added the installation cost:

$$y = 1,100x + 5,000$$

What we need now is a graph. We can try to do this by hand or by using a graphing calculator. It's up to you. Either way, you have to decide the boundaries of the graph. We certainly want 5 years and 30 years showing up on the graph.

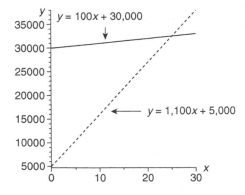

The solid line shows us the cost of the solar power system and the dotted line shows us the cost of the non-solar electrical system.

The power of the visual is that the black line is flatter than the gray line, which conveys very clearly that the cost of paying the power company for electricity is far more in the long run than the cost of solar power. Also clearly visible on a graph, which was not easy to calculate using numbers, is the point at which the two lines intersect. On the graph, that point is somewhere between 20 and 30 years.

Now we can use algebra to figure out where this intersection occurs. To construct the equation for x, we have to eliminate y. Because we need to find where the costs are the same, the two expressions for y must be equal. This gives us the algebraic equation whose solution answers our original question: At what point does solar power pay for itself?

$$100x + 30,000 = 1,100x + 5,000$$

The solution to this equation is $x = 25$. The answer is 25 years.

How might this help you in the real world? Imagine that you are building a house and have to decide which type of system you want to install. By doing this problem, you can see that if you plan on living in the house for less than 25 years, you might prefer to install the traditional electric system. Otherwise, the solar power system is cheaper. Today, the cost of installing and maintaining a solar power system may be significantly cheaper. It will likely be much less than 25 years for a solar power system to pay for itself. You may have other considerations in your decision—such as the impact on the environment and tax advantages.

It always helps after finding a solution to go back and think about how you got there. For this problem we started by exploring—trying out some costs over time.

Then we turned to algebra. Then we empowered the algebra with a graph. By going back and forth between the algebra and the graph, we achieved three goals: We constructed the equation—the hardest part; we solved the equation; and, most important, we understood what the answer told us!

What you should notice here is that no one single strategy led us from the beginning of the problem to the answer. We had to try different strategies at different times. This is what we meant at the beginning of this chapter when we wrote: There is no one right way to solve a problem. These are personal choices. Because they're personal, the strategies you choose may be different from those used by your instructor, your textbook, and your friends.

Following your own approach goes a long way toward eliminating math anxiety. You are on familiar ground. You are playing to your strengths. You are using something that makes sense to you. At the same time, you're enlarging your tool box and learning something new.

The Tire Problem Revisited

What makes some problems more difficult than others, even when the underlying mathematics is the same?

Here is the tire problem from Chapter One:

Problem: *A car is driven 5,000 miles. Its five tires (one is the spare) are rotated regularly and frequently (every few miles). How many miles will any one tire have traveled on the road by the end of the trip?*

We still need to come to terms with what the meaning of the scenario is. Rotating tires is the process of moving the tires to different wheels so that no one tire loses more tread over time than any other tire. Tire rotation is part of car maintenance, and is supposed to take place every 6,000 to 9,000 miles driven.

A picture will help explain what this means. If we label our tires a, b, c, and d, and if the box represents the car, here is a diagram of two possible tire rotations:

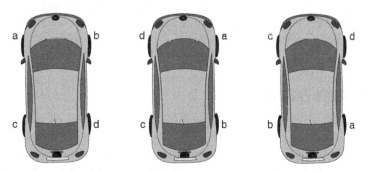

This diagram illustrates why we say that tires are "rotated": if you look at tire a, its position rotates around the car in a clockwise fashion.

In the scenario from the problem, there is a fifth tire—the spare. Often, the spare tire is used only if one of the other tires goes flat. Spare tires are smaller than normal tires and cannot handle higher speeds. In this problem, we shouldn't worry about such realistic details, and treat the spare tire like any other tire. Let us label this tire with the letter e.

We can now start with an initial strategy. One idea is to sort out the possibilities with a table. This time, the table will illustrate the sequence of tires being rotated. Then we can just add up what is in front of our eyes as follows:

Miles	0–1,000	1,000–2,000	2,000–3,000	3,000–4,000	4,000–5,000
Tires on Car	abcd	bcde	cdea	deab	eabc
Tire in Trunk	e	a	b	c	d

This layout allows us to see that each of the five tires has been in the trunk only once during the rotation modeled earlier. That means each has spent 1,000 miles in the trunk. The simplest way to calculate the number of miles that any one tire has been on the road is by subtracting the number of "trunk miles" from the total journey of 5,000 miles, giving us 4,000 miles.

Alternatively, we could create a unit of "tire miles." If the car goes a total of 5,000 miles, then four tires on the car go a total of 20,000 "tire miles." One-fifth of the total 20,000 tire miles would be the share of each tire: 4,000 miles.

The following problem sounds different, because the drivers are "rotating" instead of the tires. However, it is actually the same problem!

> **Problem:** *Five friends decide to take a 5,000-mile car trip together. They are equally competent and enthusiastic drivers and so they will take equal turns at the wheel. For how many miles will any individual not be driving?*

Look how much simpler this problem seems to be than the original "tire problem." It uses the same numbers and the same concept, but reversed. Now the "tires on the road" are the non-driving passengers, and the tire in the trunk is the driver.

The point here is that context counts. The second version of the problem presents a context in which you could easily place yourself. For the first version of the problem, can you imagine yourself as a tire? To solve the problem, additional work is needed to conceptualize what is actually happening to the tires.

Divided-Page Problem Solving

When you are problem solving, using the divided-page approach helps you keep a running record of your ideas and insights. It serves a therapeutic purpose in giving you the opportunity to purge your mind of negative thoughts about yourself, the problem, and your math class. It helps you keep track of what you are doing and why. This discipline tends to prevent some common errors, such as putting down wrong units or making careless mistakes in calculation. Finally, this technique develops your skill in listening to yourself think.

Let's see how one student's divided page gives her a way to solve a problem that is typical in college-level math courses.

Problem: Find the domain of the function $f(x) = \dfrac{x-3}{x^2-1}$

My feelings and thoughts	My work
Oh man, not one of these! I hate these problems. When will I ever need to do this in my life?	
OK, I need to focus. I know I am supposed to solve some equation. But what equation?	
Possible equations: $x - 3 = 0$, $x^2 - 1 = 0$. Which one?	
What else do I know? My teacher always tells us "division by zero is undefined" – what does this mean?	
This means either $x - 3$ or $x^2 - 1$ can't be zero. But which one?	
Well, let's say $x = 3$. Then $f(3) = 0/8$. This means $0 \div 8$. The calculator says that this is 0. I suppose that should be OK.	
What if $x^2 - 1 = 0$? Well, I know that if $x = 1$, then $x^2 - 1 = 0$.	
$f(1) = \dfrac{1-3}{0} = \dfrac{-2}{0}$. This is $-2 \div 0$. The calculator tells me that this is an error. I guess this can't be done. This must be what my teacher means about division by 0.	
The equation must be:	$x^2 - 1 \neq 0$
How can I solve this? Let's try to factor! Ah ha! I have it now!	$x^2 - 1 = (x - 1)(x + 1)$. $(x - 1)(x + 1) \neq 0$, so $x \neq 1, x \neq -1$.
How should I write this? Well, any real number besides 1 and -1 is OK. May be I can just say that!	All real numbers besides 1 and -1.

Notice how much the student wrote on the left side before beginning to write any work on the problem on the right! After the initial reaction typical of first-year college students to this type of problem, she became quite relaxed about playing with the problem and reasoning out what she couldn't remember. She knows that trying something out, even if it doesn't end up contributing to her solution to the problem, will be helpful and get her "unstuck." For example, when she couldn't remember what "division by zero is undefined" meant, she tried to compute $0/8$ and $-2/0$, constantly "unpacking" the notation. Ultimately, this led to a solution to the problem.

Problem Solving for Online Homework

How can you adopt the techniques from this chapter to online homework? You should always be *writing* when doing math. This does not change when you are doing your problems online. Keep paper and pencil with you when you are at the computer. You can even continue to use the divided-page technique. When you are getting help doing a problem, stay one step ahead of what's on the screen. Write down why each step is done. Tinker with the problem—figure out how the answer is altered when you change something.

As with paper-and-pencil assignments, it is important to manage your time. You need as much time for online problems as you need for paper-and-pencil problems. Schedule time during your day when you plan on working on your homework, and try to do so in a quiet location where you can concentrate.

Advantages and disadvantages of online problem solving

There are several advantages to having your homework assigned online. First, you get immediate feedback. When you enter your answer to a given problem, you know immediately if you are right or wrong. Second, the software may provide options to display a portion of the text or some other helpful hint. Most software will allow several attempts on a problem so that if you do it incorrectly once, you can try again! Another common online feature is that you can e-mail your instructor directly from a problem when you are stuck. The e-mail will contain a link directly to the problem. You can do this at any time of day. You may not get an answer from your instructor in the middle of the night, but you can send it when you have a question.

In addition to communicating with your instructor, the program itself can provide you with personal attention by way of diagnostic testing. These tests may be optional, but they will let *you* know what you already understand and what you need to work on. The software might even formulate a study plan to help you turn your biggest weakness into your biggest strength.

However, online homework programs have some disadvantages. One problem that sometimes arises is with language. You may type a correct answer to an algebra problem as $1 + 1/x$, but the computer is expecting $(x + 1)/x$, an equivalent answer. If you are absolutely convinced that your answer is correct, try to find alternative algebraic forms. This is a good way to exercise your algebra skills. This issue is becoming a less serious problem as many online mathematics programs are being programmed to recognize equivalent expressions.

The biggest pitfall when doing online homework is assuming that you did everything correctly if the computer tells you that your answer is correct. Computers can only grade your final answer and don't see your work, which is what your instructor probably cares about the most.

To illustrate, suppose you are asked to simplify the following logarithmic expression:

$$\ln\left(\frac{xy}{z}\right)$$

You write on your scratch paper:

$$\ln\left(\frac{xy}{z}\right) = \frac{\ln(xy)}{\ln(z)} = \ln(xy) - \ln(z) = \ln(x) + \ln(y) - \ln(z)$$

You type your final answer into the box on your computer screen, and a nice green check mark pops up telling you that your answer is correct. Sounds great, right? When you take your exam the following week, you see the same problem. Because you did it correctly before, you confidently rewrite your solution. Because your instructor makes you show all of your work, you write in the steps you wrote before. When you receive your graded test, this problem is marked 3/10 and the first two equal signs have big red X's. What happened?

In this situation, what happened is that you made two errors in your reasoning that undid one another, allowing you to arrive at the correct answer. This is the mathematical version of "two wrongs make a right." Yes, you obtained the correct answer. However, college mathematics instructors are at least as interested in your logic and your ability to reason through the problems as they are in the final answer. This situation is particularly frustrating because the computer told you that you were correct, and your misconception was not pointed out until your graded exam was returned!

The solution to this problem lies in you. Give yourself permission to question your reasoning as well as that of your study partners. Make sure you are confident in every step of your solution to a problem, and don't be afraid of doubt! For every doubt you have, there is a teachable moment waiting for you and your classmates. Do not be afraid to ask your instructor about these doubts in class. Odds are that several of your fellow classmates have the same question! You should be willing to visit your instructor during office hours to ask about these questions or discuss them with your study group. Courses that use online homework give you the opportunity to be proactive about the way you learn mathematics. Seize the opportunity and you will probably find out that you understand math better than you ever expected!

Finally, it is important to prepare for tests by practicing problems without the instantaneous feedback provided by online homework systems. When you are taking your exam, you will not receive feedback after you complete each problem. Find ways to boost your confidence without being signaled that your answer is correct. One way to do this is to check over your work. You can also see if your answer is a *reasonable* solution to the problem. You may also find a way of double checking your result on a calculator.

Practicing error analysis

Another issue that arises with online homework occurs when your final answer is marked incorrect. You may be tempted to change your answer without analyzing what may have gone wrong. Don't. Try to do an *error analysis*. You can have the same opportunity when you check the answers in the back of your textbook and find out your answer is wrong. Don't just change your answer. Your answer may be correct, but in a different form. If you have made an error, this is an opportunity to learn! Give yourself the time it takes to go through this process carefully. The benefit will be enormous.

Where do we begin? Reread the problem. Ask yourself whether the answer should be a number, an algebraic expression, a graph, or something else. What variables, if any, should be involved in your answer? For example, suppose the problem asks you to write a formula for the area of a circle in terms of its circumference, c. If you wrote $A = \dfrac{cr}{2}$, your error is that the only variable that is supposed to be on one side of your equation is c. You had another variable, r.

Next, read through every step in your work. You may find an error in your reasoning. Perhaps you stopped too soon. Perhaps you miscopied one line to another. Perhaps you didn't expand or factor an expression correctly. Perhaps you dropped a minus sign somewhere. Examining your work will allow you to find these errors.

Finally, check to see if you understand how the pieces of your problem fit together. It could be that your approach is wrong because of something you misunderstood.

One key ingredient in the error analysis we are recommending is that you write your work down on paper as you work through the problems. You won't be able to identify doubts or misconceptions if you haven't been writing. And you won't be able to check your work.

Problem Solving on Tests

What about on those tests? On tests, you won't have time to try all of the different strategies we used in the solar power problem. It is not realistic to use divided-page techniques during an exam. However, if you solve enough problems when you are studying, your experience will help you select more efficient strategies for problems on tests.

What if you handle your math class, math homework, and your test preparation without experiencing moderate or even intense anxiety and you are still not doing as well in college math as you have to? Consider the possibility that you don't have math anxiety. Instead, you might have test anxiety, a condition that is commonly confused with math anxiety. This is because math tests come so frequently, are so exacting, and grades aren't usually assigned on a curve.

Many colleges have test-anxiety workshops and support services. Check these out. The Internet features a whole bank of test-anxiety "treatments." For example, one way to handle test anxiety is to arrive to your exam about five minutes early and write on a blank sheet of paper everything you are worried or stressed about. Then wad your paper up and throw it away.

One source of test anxiety comes from worrying about how your performance on this exam, or even a specific problem, impacts your course grade, your GPA, your eligibility for school sports, and even your financial aid! This writing exercise allows you to purge all of these concerns so they aren't cluttering your mental pathways and preventing you from using your problem-solving skills.

Another possible source of test anxiety may come from worrying about whether or not you will remember certain facts that you may need, such as the quadratic formula. We have a suggestion to handle this. When you receive your blank test, or a sheet of scratch paper, from your instructor, write down everything that you have cluttering your short-term memory. If it comes into the test in your brain, it isn't cheating! If you do this, the pathways in your brain will be free to function as if you were working at home.

Our favorite approach is that of the "Praxis" group, part of the Educational Testing Service. Based on the Praxis model, here are some alternate ways to think about a test that may decrease your test anxiety.

Perfectionism	Realism
There is an impossible amount of things to learn for this test. My knowledge of _____ is shaky.	I don't need to know every question to pass. If I start studying now, I can learn what I need to know.
Negative	Positive
I always do poorly on tests. If I don't pass, this test, I'm a failure. This test is going to have a trick question.	I've got a better study plan for this test than I ever had before. I'm going to pass but if I don't, I can bounce back. This test is designed to let me show what I know.

The left column displays self-defeating statements. Whenever one of these thoughts occurs to you, replace that thought with its counterpart in the right column. If you aren't taking the exam at the time, you might even say the counterpart on the right out loud.

The Fine Art of Problem Solving

What makes for good problem solving? Of course, familiarity with mathematics is important, but so are patience, persistence, practice, and experience. It may seem that math teachers, who are experts, are just pulling the problem-solving steps out of the air. In fact, ideas are coming from all of their past experience and close scrutiny of the problem. A few of the many things you can do to tackle a problem are:

1. Keep track of your thoughts, as with the divided-page technique.
2. Pursue all possibilities.
3. Line up your data in tables and look for patterns.
4. Plot graphs to help see relationships.

Skill comes from experience, and, as you will find when you banish math anxiety, confidence comes from both.

Short Take 5

Studying in a Group

It used to be that talking to your neighbor in a math class was unacceptable to the teacher, even if you weren't taking a test; but certainly conversation during a test was interpreted to be cheating. That prejudice carried over to doing homework. If you did homework with a classmate, wouldn't that be cheating, too? Especially with standardized tests dominating the curriculum, math problems would have a single right answer and if two or more of you came up consistently with the same answer and similar worksheets, you'd be considered not to have done your own work.

The truth is that studying math in a group is the most effective way to learn and retain what you've learned. So in this short take, we want to persuade you that studying in a group is worth doing and explain how to structure a group that will serve everybody's needs.

Who should I study with?

Choosing your study-mates is really important. It is not as important as choosing a spouse. But you need to find individuals who will work well together as a group—who respect your need to be listened to and aren't going to exploit one another. Your first instinct might be to choose a high-performing math student as a group member or group leader. But you need to feel both that you can contribute and that you can benefit. It's a balance. You will also want to be able to change group members if the initial selection doesn't work out.

Ground Rules: How do we work together without an instructor supervising our work?

One thing is to start very early in the semester, as early as possible. Set a regular meeting time and stick to it even if some individuals have to be absent some of the time.

Another is to set reasonable expectations and specific ground rules. For example, no putting down of a member of the group for getting a wrong, even an absurdly wrong, answer is allowed. No competition in class or in the group is allowed. Everyone will contribute productively; no one just takes without giving, and "giving" might involve assigning specific problems to individuals in the group, so that when the group meets, some members are more prepared than others—but not all of the time, and not always the same members. An important ground rule is not to make fun of a group member's mistakes or slowness to grasp a concept.

Structuring the study group's time per individual session

What follows could be a one-hour session.

Time for review of textbook and class notes—10 minutes

Time assigned to a single problem in homework; one person presents, or everybody talks at once; one person goes to the white board and takes notes from what the others tell him or her to write down—30 minutes.

Time for reflection on the problem after it's solved: what made it difficult for the group? For any one individual? This exercise forces you to pay attention to the difficulties and strategies of the others in the group and allows you to notice what was "hard" and how to solve similar problems—20 minutes.

John, Joan, Susan, Marco

John: Hi gang. Glad you're all here. Well, what was this week all about? Seems to me we were going over problems with one variable.

Joan: Yeah, that first problem was

$$x + 38 = 7x + 2$$

[Susan goes to the white board to write down the problem so everyone can see what it looks like. Another advantage to having a white board hanging is that everyone can see the problem and see and listen to one another at the same time.]

Marco: The goal is to find out the value of x. And we've got x on both sides of the equals sign, so that should be easy.

Susan: Shall we start with the rule? Or first look at the two sides of the equation? Is there anything to notice about this particular equation before we apply the rule?

John: Gee. Some number plus 36 equals six times that number plus 2. I don't think x is very large. What do you think Joan?

Joan: OK. Let's try a number and see. Let's take 2, x equals 2.

John: So $2 + 38 = 14 + 2 = 16$? That's not going to work. But what have we learned about x?

Marco: x has to be a lot bigger than 2, twice 2 or more.

Susan: Now let's apply the rule. The rule is: Gather like terms. Collect all the x's on one side of the equation. [She writes on the board]

$$x + (-7x) = -38 + 2$$

It looks funny but it's right. Isn't it? All the *x*'s are on the left side; all the numbers on the right. Now we have to simplify. [Susan is still writing.]

Marco: *That's −6x on the left-hand side and −36 on the right.*

Joan: *Why is it minus 36? What's confusing me is that we have a number with a minus sign and a number with a plus sign that we're adding. Is that OK?*

Marco: *Let's not stop here. Let's continue getting a value for x and see if that solves the equation.*

Susan: *In −6x = −36, the minuses cancel out; x = 6.*

The group continues talking even after they get an answer.

Does 6 make sense?

Could there be another answer?

Let's review adding and subtracting negative numbers.

Chapter Seven

From School Mathematics to the Workplace

The purpose of this chapter is to expose you to real-world math, especially the kinds of problems you may encounter in any employment sector. If we're successful, this chapter will leave you with the most useful skill in real-world mathematics: the confidence to figure things out for yourself.

Your real-world use of mathematics depends, of course, on the math you learn in school. But rarely will the math-related problems you have to solve when you get into the real world of work look exactly like the ones you've solved in class.

So how are you going to prepare?

Pre-Employment Testing

Pre-employment tests are tests of your readiness for employment. You will find them in several empxloyment sectors, such as the military, law enforcement, civil service, health care, and banking.

Pre-employment tests look very much like the kind of math you are expected to have mastered in school. In addition to the scope of your knowledge, the employer is testing how well you can follow instructions and how well you can apply what you know. A successful applicant will be able to:

1. *Recognize* what the problem calls for,
2. *Recall* any necessary facts, and
3. *Apply* what he or she knows to a new kind of challenge.

Test questions

Here's a typical question from a study guide for a generic civil service exam:

A recent poll showed that 42 percent of people polled favored a new congressional bill; 28 percent were opposed to it; 20 percent were neither for nor against it. And the rest did not vote at all. What fractional part of the poll did not vote at all?

Answers given: 1/10, 1/9, 2/10, 9/10[1]

---

[1] Kaplan Civil Service Exam Second Edition, 2008, p. 209.

The first thing to notice is that, as a percentage problem, the total should add up to 100 percent, so you should start by adding 42% + 28% + 20% = 90%

Now you should return to the question and notice you're being asked to give an answer in fractions, not percents. Be careful! You are being asked specifically for the *fractional part of the poll* who did NOT vote at all. So 90 percent is not the answer. Nor is 9/10 the answer. As a percentage, the answer is the difference between 90 percent and 100 percent, which is 10 percent. The fraction equivalent to 10 percent is 1/10.

In this question, you should *recognize* that you are asked about percents and fractions.

This problem requires that you *recall* that percents and fractions are stand-ins for one another. You are supposed to remember how to convert one into the other.

Finally, you have to *apply* what you remember to this particular problem. This involves reading it carefully, reading it a second time, and returning to the problem to test whether the answer makes sense.

Let's try another question from a prep course for a police force's qualifying exam:

During a typical work week, Officer Munez drives his police cruiser approximately 1,500 miles, which is about 300 miles/day. Officer Munez called in sick on Tuesday and only worked four days this week. About how many miles did Officer Munez drive this week?

Answers given: 1,000 miles, 1,200 miles, 1,500 miles, 1,800 miles[2]

One way to start working on this problem is to eliminate the obviously wrong answers. If Munez is sick one day, then he has to have driven fewer miles than 1,800 (which he never drives), so 1,800 is wrong. He can't have driven 1,500 miles because that's what he drives in a regular five-day work week. So the answer is either 1,000 miles or 1,200 miles.

To figure out which of these two is the correct mileage, you only have to subtract his average daily mileage of 300 miles from his weekly average of 1,500 miles to get to 1,200 miles.

- *Recognize* that this problem involves averages and subtraction.
- *Recall* how to do the necessary arithmetic without the aid of a calculator.
- *Apply* simple arithmetic—division and subtraction with numbers larger than 1,000.

In addition to "arithmetic reasoning," pre-employment exams also test "mathematics knowledge." Here's typical question from an aptitude test for military service used for both entrance and assignment. The test matters a lot because the final score will determine whether the applicant is qualified for education and training.

---

[2] Cliff Notes, *Police Officer Exam Cram Plan*, Wiley, p. 177.

If a rectangular solid has a length of 10 meters, a height of 5 meters, and a width of 3 meters, what is its total surface area?

Answers given: 30 square meters, 60 square meters, 150 square meters, 190 square meters[3]

What you have to know to answer this question is the meaning of the term *surface area*, and, even better, to be able to picture in your mind a rectangular solid in three dimensions, compared to a rectangle that is flat.

In a geometry problem such as this, the most helpful problem-solving strategy to get you started is to *draw a picture*, and then label it with the information given in the problem:

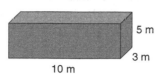

You are being asked to calculate "surface area." This involves adding the areas of every side, or *face*. On a rectangular solid, every face is rectangle. To do this, first you compute the area of each of the faces. You will make the calculation easier if you remember that opposite faces in a rectangular solid have the same area. So the strategy is: compute the areas of the three faces and double the total.

$$2 \times (3 \times 10 + 10 \times 5 + 3 \times 5) = 2 \times (30 + 50 + 15) = 190,$$
a total of 190 square meters.

- *Recognize* the nature of a three-dimensional block and the meaning of the term *surface area*.
- *Recall* how to compute the surface area of a figure such as a rectangular block.
- *Apply* this knowledge by performing the computation.

Pre-employments tests divide the test questions into "arithmetic reasoning" and "mathematics knowledge." The mathematics knowledge in the last question about area is a good example. To solve the problem it is not enough to do the arithmetic correctly. The test taker has to understand and *recall* how the area of a solid is calculated.

Here's another multi-step problem:

If a person earns $200 for a regular 40-hour work week, and overtime is calculated at 1.5 times the regular rate for any additional hours, how many hours must be worked for that person to earn $230 in the week?

Answers given: 43 hours, 44, hours, 45, hours, 46, hours[4]

---

[3] Laura Stradley and Robin Kavanagh, *The Complete Idiot's Guide to ASVAB*, Alpha Books, 2010, p. 213.
[4] Adapted from the Kaplan Civil Service Exams, 2nd ed., 2008, p. 208.

The first task for solving this problem is to calculate the regular hourly rate by dividing $200 by 40 hours. This gives you a wage of $5/hour. At time-and-a-half, the overtime rate is $7.50.

To solve the problem, you have to notice that the $230 total earnings is made up of two components. This first is the $200 for 40 hours worked at the regular, non-overtime wage. This leaves $30 of overtime pay. Divide $30 by $7.50 to find out that the person worked 4 overtime hours. Finally, add the overtime hours to the regular hours to get the final answer of 44 hours.

- *Recognize* that the total salary comes from two portions, one at regular pay and the other at the overtime rate.
- *Recall* how to do those computations. This requires an understanding that the $200 total pay must be divided by the total number of hours worked.
- *Apply* what you know by putting all the calculations together.

Types of tests

Paper-and-pencil tests may be fairly familiar to you. However, in order to improve the accuracy and speed of scoring, many pre-employment tests today are delivered and expected to be taken online. If you're comfortable with online assignments, such an exam will not be particularly threatening.

What you really need to consider before you select a test-taking strategy is how the scoring will be done. Try to find out in advance whether wrong answers will simply not be counted or will be deducted from your total score.

The hardest exams to deal with psychologically are the so-called *computer adaptive exams*. These work in the following way. Your performance on one test question will determine the difficulty level of the next question the computer will deliver for you to answer. This means that each test is personalized to the individual. Scoring is based on (1) the number of questions answered correctly, *and* (2) the difficulty level of those questions.

Bottom line: be sure you read the small print before you start to work on an exam. Computer adaptive exams can penalize you for guessing. Exams that score just the number of correct answers reward good guesses. Be sure to know which scoring system is being used.

Using Mathematics on the Job

You are unlikely to encounter textbook-style fraction problems on the job. However, you will encounter quantitative scenarios in most professional employment.

For example, if you are working in the health care field, you will have to manage ratios, proportions, and rates. If you don't believe us, check out the following terms on the Internet:

1. Blood pressure
2. Pulse rate

3. EKG
4. Telemetry
5. Unit-monitoring machines
6. Unit conversion (between the metric and the English systems)

Particularly, if you work in a lab, you will have to be familiar with unit conversions and orders of magnitude.

If you enter the building trades, or become a real estate agent, you'll need to be comfortable with the language of length and area. This includes angles, areas, surface areas, volume, distribution of forces, and density. Areas are particular important. For example, rent for commercial office space is measured in dollars per *square foot*.

Of course, anyone involved in auto mechanics has to be comfortable with math. The same is true for anyone involved in auto sales. Car salespeople have to understand depreciation, especially when calculating the trade-in value of a customer's current car.

The same is true of auto parts. Anyone selling tires, in order to describe the advantage of one type of tire over another, should understand the significance of "tire pressure." Pressure is measured by a fraction; its units are pounds per square inch (abbreviated psi).

An electrician, or somebody who sets up wireless Internet systems in a building, will need to be familiar with quantities associated with circuits, such as resistance, current, impedance, and current voltage, particularly Ohm's law, which states that the voltage difference across a circuit is proportional to the current.

It is also useful to know that that *power* can be expressed as a fraction: energy/time. Power is measured in *kilowatts*. The electricity you consume from your power company is measured in *kilowatt-hours*. Think about what this means mathematically. If kilowatts represents energy divided by time, multiplying this by time means that *kilowatt-hours* is a unit of energy—the energy you consumed over the billing period.

Now you might be thinking: "I'm not aiming to be in any of those businesses. I'm heading for the culinary arts." Restaurant work, from cooking to managing, is also dependent on calculations. In cooking, you will have to manage nutrition statistics, ingredient proportions, and unit conversions such as gallons to ounces.

Cooking time may or may not vary with temperature. For example, is it okay to shorten the time to bake a cake by doubling the temperature? Better check with the chef on this one. What if *you* are the chef? Well, you should understand that the time required to bake a cake is, in mathematics language, a *nonlinear* function of temperature. In other words, proportional reasoning won't work here!

What about working in fast food? If you are a manager at a fast-food restaurant, you will have to balance the accounts at the end of the day. You may be expected to compute something called "fry efficiency." Although your paperwork will tell you a formula to compute your daily fry efficiency, and a calculator will do the arithmetic for you, it is up to you to interpret the meaning. If you understand what the formula tells you, you can determine whether improving your fry efficiency is necessary and, if so, how to do so.

As a profession, business is completely quantitative. Let's say you're asked to construct your company's market survey, where some "sample" of actual or potential customers is asked for their preferences. You'll need to know what size sample is "legitimate" for the survey as well as how to compute and interpret "margins of error." Margins of error are particularly important when you report to your boss the degree to which your results generalize. The interpretation of margins of error is treated in a little more detail later in the chapter. When presenting your findings, which are likely to include complicated numerical relationships, to your colleagues, you will find graphing to be your best friend.

The hottest new area in market analysis is called *business intelligence*. This involves something called *data mining*. Let's say your company has initiated a website to advertise its products and wants to analyze the "traffic" on the website. One question you can ask is how many new hits your website is receiving each day. How many new hits are from new people who never before logged on? How many actual sales are generated from online visitors? Can we break this information down by gender, by age, by region? Over time, how do these patterns change? Although there is no shortage of services available to collect this data for you, the analysis required to answer these questions will require *your* mathematical knowledge.

Real-World Quantitative Explorations

Depreciation

If you are in business, you will be concerned with how much your assets (such as buildings or equipment) are worth over time. Even if you are not in business, you will care about how much to save each year to replace something like your car or your refrigerator. The value of equipment decreases over time, and this loss in value is called *depreciation*. How can you measure and mathematically model depreciation?

The most common method to measure depreciation is called *straight-line* depreciation. The assumption is that the value loses the same amount every year. If you have a vehicle that you purchased for $20,000 and it loses $2,000 of value per year, then you can apply straight-line depreciation.

In this example, we can write an equation that describes your car's value as a function of time and graph it. Typically, we will want to know that if the car is, say, 5 years old, how much it is worth. So we let x, our independent variable, be the number of years since the vehicle was purchased. Then y is the value of the vehicle. The car starts out being worth $20,000, and after x years, the value is $2,000x$ less. The equation is:

$$y = 20,000 - 2,000x$$

Let's graph this equation. If you are using your graphing calculator, don't forget to specify a window!

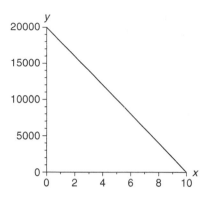

Straight-line depreciation gets its name from the fact that you have assumed the value of the car is a linear function of time and the graph is a straight line.

Notice that the *x*-intercept occurs at $x = 10$. This is the solution to the equation $20{,}000 - 2{,}000x = 0$. This tells you that after 10 years, the car has no value. In business language, the *lifetime* of the car is 10 years.

There are other ways to measure depreciation besides the straight-line method. Businesspeople will choose their method based on the tax consequences.

Margins of error

In any statistical study, you must deal with the *margin of error*. Let's say you are asked to conduct a study to determine the best price for selling a new vacuum cleaner. Part of your study involves asking participants whether they would purchase the product for $150. At the end, you compute the percentage of the sample that says "yes."

There is a *true* percent for your market, but it is too expensive to survey everyone. You have to select a smaller sample. The question is, how far off is the answer you get in your survey from the true percentage?

Let's say you find that 50 percent of respondents would purchase your vacuum cleaner. That sounds like good news! If you report that to your boss, she will want to know how confident you can be in your result. This is what the margin of error tells you.

When thinking about the margin of error, it is helpful to have a picture:

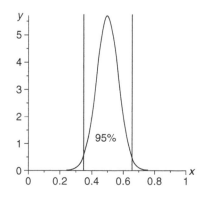

This picture tells you that with a 95 percent "confidence level," the true percentage of your market that would purchase the vacuum cleaner at $150 is between 34 percent and 67 percent. This range is called the *confidence interval*.

That is actually a pretty wide band, and your boss may want you to repeat the study with a larger sample in order to tighten up that interval.

Sampling and margins of error are examples of mathematical concepts that will be necessary in many, if not most, occupations. When you have banished math anxiety from your life, you will find yourself at ease if you have to do math at work.

Chapter Eight

Mathematics as Empowerment

We have to deal with data and quantitative information every day. When you develop yourself as a resource for understanding math, you will be able to process this information. This will empower you to make better decisions regarding the direction of your own destiny, whether through individual financial decisions or through participating in the democratic process.

Health Care Information

Math is ever present in health care, not just for the health care professional but also for the patient. You may be told that the drug you've been prescribed has a 0.1 percent chance of contributing to heart disease. How do you make sense of this information? This requires number sense and an understanding of probability.

Managing Debt

Interest on debt can be thought of as "rent" paid for the use of someone else's money.

Suppose you made $1,000 of purchases on a credit card that (a) carries a 24 percent annual interest rate that is compounded monthly and (b) allows a minimum payment each month of $10. Assuming that you do not make any additional purchases, how long will it take you to pay off your credit card?

At first glance, you may not understand the words *compounded monthly* in the problem. "Compounding monthly" is the bank's way of telling you how often interest on the $1,000 is added to your debt, but what does that mean? To find out, let's dig a little bit into how credit cards work. This is essential in order to understand this problem and its impact on your life.

If you don't pay all the interest in the month in which it's due, *interest is charged on that interest*. Effectively, the credit card company has made you an additional loan of the unpaid interest.

How is the interest on the interest calculated?

The annual interest rate is 24 percent, and this is spread evenly over the twelve months of the year. Therefore, each month the bank adds 2 percent of the balance. Your balance at the end of the first month is $1,020; this includes the $1,000 outstanding and the $20 of interest. Now you get your bill, which reads: "Your minimum payment is $10.00." You pay the minimum payment and you think all is well.

But . . . next month you get a bill with the balance $1,030.20. How is that possible? After making the minimum payment, your balance was $1,010, and this amount was assessed at 2 percent interest which added $20.20 to your balance.

You were not only charged 2 percent on the original $1,000, but also on the unpaid interest of $10. This interest will be charged on your balance at the end of every month, and that balance includes unpaid interest. This is what is meant by the statement that the interest *compounds monthly*.

Let's say you make your $10 monthly payment every month. What is your balance at the end of a year? We continue the same computations as before, rounding as necessary. This gives us the following table:

Month	Initial Balance	+Interest	−Min. Payment	Final Balance
1	$1,000	$20	$10	$1,010
2	$1,010	$20.20	$10	$1,020.20
3	$1,020.20	$20.40	$10	$1,030.60
4	$1,030.60	$20.61	$10	$1,041.21
5	$1,041.21	$20.82	$10	$1,052.03
6	$1,052.03	$21.04	$10	$1,063.07
7	$1,063.07	$21.26	$10	$1,074.33
8	$1,074.33	$21.49	$10	$1,085.82
9	$1,085.82	$21.72	$10	$1,097.54
10	$1,097.54	$21.95	$10	$1,109.49
11	$1,109.49	$22.19	$10	$1,121.68
12	$1,121.68	$22.43	$10	$1,134.11

By now this data might suggest an answer to our original question. What is happening each month? Our balance is getting larger and larger, instead of decreasing, as we thought it would. *The reason for this is that our minimum payment is less than the interest.* In fact, our interest is increasing from month to month as well! That is the effect of compounding frequency.

At present, it is illegal for a credit card company to charge a monthly payment that is less than the interest. However, laws often change, and it is valuable to understand the financial implications if this consumer protection is ever taken away.

The table also tells us that after one year, what you've paid for borrowing the original $1,000 is $1,134.11. The difference is the price you paid for the convenience of borrowing $1,000 twelve months before. If you pay off the credit card at the end of this twelve-month period, you will have paid a total of 134 percent of the original price of the goods and services that you purchased!

There are other conditions that may increase your credit card balance as well. Suppose that you had a $1,100 credit limit. By the end of the tenth month, you will have exceeded your credit limit *without having made a single additional purchase!* Once you pass the limit, the bank will charge additional fees, and those fees themselves will become part of the unpaid balance on which the interest is calculated. You should keep all of this in mind as you consider all of the credit card offers that you receive in the mail!

Compound interest is modeled with *exponential* functions. In our example, we can look far into the future and sketch a graph:

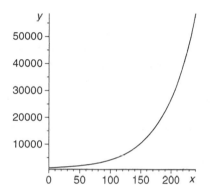

In this graph, the independent variable *x* is time, measured in months. The dependent variable *y* is the balance of the credit card. If you continue only paying the minimum monthly payment of $10, after ten years the balance is close to $60,000! While this is surely an exaggeration, the point is that you should always pay at least the total interest due on your credit cards, preferably more.

Why does compound interest result in a balance that grows exponentially? We can see this more easily if we eliminate the monthly payments. We have a debt of $1,000 and interest is accrued at a monthly rate of 2 percent, compounded monthly. How much do we owe after the first month? We owe the original $1,000 plus 2 percent of that balance:

Balance after 1 month $= \$1,000 + \$1,000(.02) = \$1,000(1.02)$

What is our balance after two months? We owe the previous balance of $1,000(1.02) PLUS 2 percent of **that** balance:

Balance after 2 months $= \$1,000(1.02) + \$1,000(1.02)(.02) = \$1,000(1.02)^2$

Balance after 3 months $= \$1000(1.02)^3$

Balance after 4 months $= \$1000(1.02)^4$

We can see the pattern. The balance, *B*, after *x* months is $\$1000(1.02)^x$. Compound interest is best modeled by an exponential function.

What role does "compounded monthly" play in all of this? Going back to the table, the first column is measured in months. But imagine that interest is compounded daily. Then the first column would be measured in days. This is actually the way interest on car loans is computed. See for yourself. Read the fine print in your car loan agreement. Once you've done that, you will see why most people, if they can, try to make their monthly car payment on the first day of the interest cycle. The moral of this story is that you should always pay attention to both the interest rate *and* the compounding frequency whenever you borrow money. This is how mathematical knowledge empowers you to manage your debt.

Credit Scores

Anyone who has applied for a credit card or a car loan knows from experience the importance of his or her credit score. When you apply for a loan, you're asking the lender to invest in you. The lender uses your credit score as part of its *risk assessment*. Your credit score determines whether or not a lender will make a loan to you and, if so, what the interest rate will be. But do you know how your credit score is calculated? In case you want to challenge that score, you should.

The most common credit score used is FICO, which stands for Fair Isaac Corporation. Your FICO score ranges between 300 and 850. The higher the score, the better your credit and the better the terms you will get from lenders.

If you inquire about the "algorithm" used to compute your FICO score, you'll find that the elements, and their weights in the computation, are as follows:

- How timely you are in paying your bills (35 percent)
- The amount of money that you owe and how much credit is available to you (30 percent)
- The length of your credit history (15 percent)
- The variety of credit in your credit history (10 percent)
- The number of recent credit applications (10 percent)

Your credit score is determined by these components and their weights. This should guide you to behavior that can raise your score.

Computing the "Value" of an Education

Any parent or educator will tell you that the value of an education can't be calculated in numbers alone. Colleges make the case for the value of an education by comparing the lifetime accumulated earnings of a college graduate to the lifetime accumulated earnings of someone who doesn't have a college degree.

Having read most of this book, you will realize that there's a lot of guessing involved in calculating these figures and that colleges don't take into account individual differences, location, and college major.

Nevertheless, you won't be able to make an informed judgment about how much college debt you're willing to carry if you're unwilling to find the numbers on which the generalizations are based.

Understanding the Income Tax System

In the United States, you frequently hear about *marginal tax rates* and sometimes you hear about *flat-tax proposals*. What do these mean?

Governments collect taxes from their citizens in order to pay for their expenses, such as maintaining an army and constructing roads and bridges. Most taxes come from an income tax—a percent of income paid by citizens to the government. In the United States, at the time this book is being written, we have a *progressive* tax system. This means, roughly speaking, that those who earn more pay a higher percent of their income in taxes.

Let's look at a simplified version of this type of system first. A salary of $100,000 sounds like a fairly large income. So we will say that those who earn less than $100,000 will pay a lower percentage than those earning more than $100,000 per year. We will set the lower percentage to be 10 percent. The higher percentage will be 25 percent.

The way this will work is that somebody who earns more than $100,000 in a year will pay the lower percentage, 10 percent, on the "first $100,000" of income and pay the 25 percent rate on the income over $100,000.

We can write algebraic expressions for these different scenarios. Because we usually know our income, that will be the *independent* variable, x, and the taxes due will be the dependent variable, y. To keep the numbers manageable, x will be measured in thousands of dollars.

There are two possibilities. If we earn less than $100,000, our taxes are 10 percent of our taxable income. In terms of our variables, if $x \leq 100$, $y = 0.1x$ (because 10 percent of x is equivalent to 0.1 *times* x). On the other hand, suppose we earn more than $100,000. In this case, $x > 100$. We pay 10 percent of our first $100,000, which is $10,000. In our units of thousand dollars, this is 10. We then add 25 percent of our income over $100,000. In our units, this additional income is $x - 100$. Therefore, in this "income bracket" our taxes are given by the formula $y = 10 + 0.25(x - 100)$.

This is another example of a piecewise-defined formula (or function). A graph of this equation is:

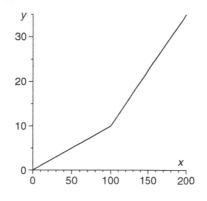

In mathematician's notation, we write:

$$y = \begin{cases} 0.1x, & x \leq 100 \\ 10 + 0.25(x - 100), & x > 100 \end{cases}$$

You can see that on one "piece" of the possible x-values, between 0 and 100, you have a line with a slope of 0.1. On the other piece, you have a line with a slope of 0.25. These slopes are the tax rates for income in the two brackets, called the *marginal tax rates*.

Let's now look at our current system. There are different types of taxpayers, so your rates and brackets depend on whether or not you are married, for example. Let's assume you are single. Then, for the year 2011, the income tax rates are:

Annual Income	Marginal Tax Rate
Less than $8,500	10%
Between $8,501 and $34,500	15%
Between $34,501 and $83,600	25%
Between $83,601 and $174,400	28%
Between $174,401 and $379,150	33%
Greater than $379,151	35%

Let's say you earn $100,000. Then your income can be broken into the brackets leading up to $100,000. We compute the taxes on the income in each bracket and add up the total.

Bracket	Income in Bracket	Marginal Tax Rate	Taxes
$0 to $8,500	$8,500	10%	$850
$8,501 to $34,500	$26,000	15%	$3,900
$34,501 to $83,600	$49,100	25%	$12,275
$83,601 to $174,400	$16,400	28%	$4,592
Total	$100,000	N/A	$21,617

Let's graph the tax function, with x being income (again, in units of thousand dollars) and y being taxes due:

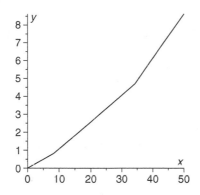

We plot this function for incomes up to $50,000. This shows the first three tax brackets. If you look closely, you can see that the graph gets steeper—the slope increases—as you pass from one bracket to the next.

In policy discussions, debates typically center on the marginal tax rates and the boundaries of the tax brackets. One common proposal heard every four years (at election time) is a flat-tax proposal. A flat-tax system is one in which every taxpayer pays the same tax rate regardless of income.

Here is a graph of a 20 percent flat-tax system:

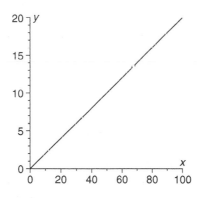

What is so flat about the flat tax? The line itself is "flat" in the sense that it is perfectly straight. But, it is not "flat" in a mathematical sense because it is not a horizontal line.

Progressive and flat-tax systems both have advantages and disadvantages. Our goal is not to persuade you one way or another. Rather, we want to empower you with the quantitative tools to *understand* the debate in order to make an educated judgment at the ballot box.

Sports Statistics—The Math Behind *Moneyball*

Statistics and numerical data are very important to sportscasters and fans. In football, a running back has rushing statistics. In basketball, every player has a free-throw percentage. In baseball, players have a batting average. What do these statistics really mean?

Let's look at the free-throw percentage for a basketball player. This statistic is computed by dividing the number of successful free-throws by the total number of free-throw attempts. This number is supposed to give the viewer (and the coach of the opposing team) a clue as to whether the player will sink his (or her) next free-throw. This type of probability is called *experimental probability* because it's a probability based on past data (in this case, the player's record of sinking free-throws).

In addition to experimental probability, there is also *theoretical probability*. Experimental probability is derived from data. Theoretical probability is different. The fact that each side of a coin has an equal chance of facing upward or downward when tossed leads to a theoretical probability of 50 percent for either outcome. But this in no way predicts what the next toss of the coin will be.

Back to the free-throw example. Have you ever noticed that the free-throw percentage of a young basketball player without much experience will fluctuate wildly with every free-throw attempt, whereas the same overall statistic for a more established player changes very little, if at all? This is an illustration of a mathematical principle called the *Law of Large Numbers*. As mathematicians state it: As the number of trials in an experiment gets large, the experimental probability approaches the theoretical probability.

One of the most celebrated use of sports statistics was made by the management of the Oakland A's baseball team in the early 2000s (a story told in the book and movie by the same name, *Moneyball*). The team manager, aided by a computer analyst, used a different statistic to calculate the value of a player to the game. Instead of focusing on runs batted in—the typical "variable" used in calculating the value of a player—the manager focused on "on-base percentage." On-base percentage measures how often a player got onto base, whether from hitting, walking, or stealing a base.

Using careful statistical analysis, it turned out that on-base averages had more predictive power than those statistics typically used to make hiring decisions. Even though the team was perennially short of money, this analysis directed the A's to players who were undervalued by the market, at salaries the team could afford. As a result, the A's found themselves competitive with top teams such as the New York Yankees.

Conclusion

Now that you know the great truth, that mathematical rules govern phenomena from the motion of the planets to the value of a baseball player, we have one more gift to give you. It takes neither a "mathematical mind" nor a lengthy sequence of math courses to master the math you need when you need it.

If you read back over this chapter and the one before, notice how many applications we were able to address using very few mathematical concepts:

- Proportions
- Linear equations
- Slope
- Piecewise functions
- Exponential growth
- Graphs

With these tools in hand (and no anxiety about using them), you will be able to analyze asset depreciation, sales reports, income tax policy, and decisions about when and how much to borrow.

Our purpose in traveling so far and so fast has been to demonstrate that the kind of math that is within your reach not only provides access to higher-paying jobs, but can give you back control over your life. The reason is straightforward: Anyone who masters college-level math has the means to figure things out. That's *empowerment*, whether you're making a purchase, negotiating a loan, or challenging a political candidate's notion of what is an acceptable tax system. Experts, who set prices or interest and mortgage rates, and bosses, who set salaries and benefits, count on the general public *not* being able to challenge their methods or to do the calculations on their own.

When you have banished math anxiety from your life, you will be able to take them on.

THE MATH LEARNER'S BILL OF RIGHTS

- I HAVE THE RIGHT TO LEARN AT MY OWN PACE

- I HAVE THE RIGHT TO ASK QUESTIONS

- I HAVE THE RIGHT TO SAY I DON'T UNDERSTAND

- I HAVE THE RIGHT TO ASK FOR EXTRA HELP

- I HAVE THE RIGHT TO LEARN FROM MY MISTAKES

- I HAVE THE RIGHT TO EVALUATE HOW WELL MY MATH INSTRUCTORS TEACH

- I HAVE THE RIGHT TO RELAX IN MY MATH CLASS

- I HAVE THE RIGHT TO DEFINE SUCCESS IN MY OWN TERMS